D0229172

30130 149661091

ADAM HART-DAVIS

JUST ANOTHER DAY

The Science and Technology of our Everyday Lives

Thurrock
Council Libraries

© Adam Hart-Davis 2006

The right of Adam Hart-Davis to be identified as the author of this work has been asserted by him in accordance with the Copyright, Designs and Patents Act 1988.

First published in hardback in Great Britain in 2006 by Orion Books
an imprint of the Orion Publishing Group Ltd
Orion House, 5 Upper St Martin's Lane,
London WC2H 9EA

1 2 3 4 5 6 7 8 9 10

All rights reserved. Apart from any use permitted under UK copyright law, this publication may only be reproduced, stored, or transmitted, in any form, or by any means, with prior permission in writing of the publishers or, in the case of reprographic production, in accordance with the terms of licences issued by the Copyright Licensing Agency.

A CIP catalogue record for this book is available from the British Library.

ISBN-13: 978 0 75287 334 3
ISBN-10: 0 75287 334 2

Jacket design: Nick May
Text design: www.carrstudio.co.uk
Printed and bound in Spain by Cayfosa Quebecor

The Orion Publishing Group's policy is to use papers that are natural, renewable and recyclable and made from wood grown in sustainable forests. The logging and manufacturing processes are expected to conform to the environmental regulations of the country of origin.

Every effort has been made to fulfil requirements with regard to reproducing copyright material. The author and publisher will be glad to rectify any omissions at the earliest opportunity.

www.orionbooks.co.uk

CONTENTS

Acknowledgements 7

1 GETTING OUT OF BED

Biological Clocks 8
Alarm Clocks 9
The Bedside Light 18

2 WASHING AND BRUSHING UP

Exercise 24
The Shower 25
Soap, Gel and Shampoo 28
Drying Your Hair 32
Mirrors 34
Cleaning Your Teeth 36
Make-up and Moisturiser 43
Body Odour, or BO 45
Shaving 46
Shaving in History 49

3 THE OUTER AND INNER YOU

Clothes 54
Shoes and Socks 58
Keeping Time 60
Furniture 62
Breakfast 67
Visiting the Lavatory 76

4 GOING TO WORK

Home and Away 86
Cycling Adventures 87
What to Wear on the Way 93
The Development of Rubber 96
Umbrellas and Hats 97
London Transport 102
The Development of the Wheel 105
Using Commuting Time 108

5 WORKING

Always Something to do 110
Writing Things Down 111
Pencils 112
Paper 115
Typewriters 118
Computers 120
Fax Machines 126
Telephones 128
How Phones Work 131
Photography 134
Flow 141
Coffee Break 144
Radio 145
Television 149
Make-up 156
Lunch 158
Drink 163
Lecturing 168

6 AFTER WORK

Sundowner 170
Ice in the Drink 171
Dinner 174
Artificial Light 181
The Remains of the Day 184
Going to Sleep 188

Index 189

ACKNOWLEDGEMENTS

Many people have helped me with this book by suggesting ideas and topics that should or should not be included, by advising me about the science behind many of our daily materials, and by reading and criticising parts of the text. They have corrected dozens of errors, but I expect I have managed to sneak others past even their critical eyes.

My long-term friend and colleague Dr David Jones read almost the whole book, and provided pages and pages of tough constructive criticism, which greatly improved the material. I also received much advice and information from my partner, Dr Sue Blackmore, and from Professor Martin Addy, Bill Gorman at The Tea Council, Walter Haegele at Braun, Deborah Hutchinson, Dr Peter Jones, Chris Marsden at Shavers.co.uk, Kevin Powell at Gillette, Dr Suzanne Allers, Raniero Destasio, Paul Matt, Laurence Punshon, Kathy Rogerson and Nickie Wheeler at Procter & Gamble, Dr Tracey Rihll, Shamir Shah, Emily Troscianko, Jolyon Troscianko, Professor Paul Walton and Simon Welfare.

I drew heavily on Henry Petroski's 400-page book on *The Pencil* (Knopf, New York, 1999) and on my friend Stephen Halliday's *Underground to Everywhere* (Sutton, Stroud, 2001), and I am most grateful to the staff of the British Library for their skill in unearthing old patents.

Finally, my publishers, Orion Books, were immensely constructive: the idea for the book was developed in discussion with Ian Marshall, and Lorraine Baxter was unfailingly supportive. Both provided many helpful comments and replied instantly to all my questions. Thank you all.

Adam Hart-Davis
April 2006

1 Getting Out of Bed

BIOLOGICAL CLOCKS

Waking up is never easy, I know.

At home, I usually wake early and stagger out of bed, ready for a shower and some work, or a bit of exercise. This is the moment in the day at which my mood is most variable. On some days I feel full of beans, bursting with energy, even though I am still deliciously woozy with sleep. On other days I just want to go back to bed. The difference may be to do with how long I have slept, or how deeply, or some combination of the two. If I am to feel great, my body clock needs to be satisfied.

We all have built-in biological clocks that follow circadian rhythms and tell us fairly accurately what time it is. The main daily clock is a bunch of cells called the suprachiasmatic nucleus, or SCN, which sits in the middle of the brain and produces a signal every twenty-four hours.

I love my radio-controlled alarm clock – except when it wakes me up.

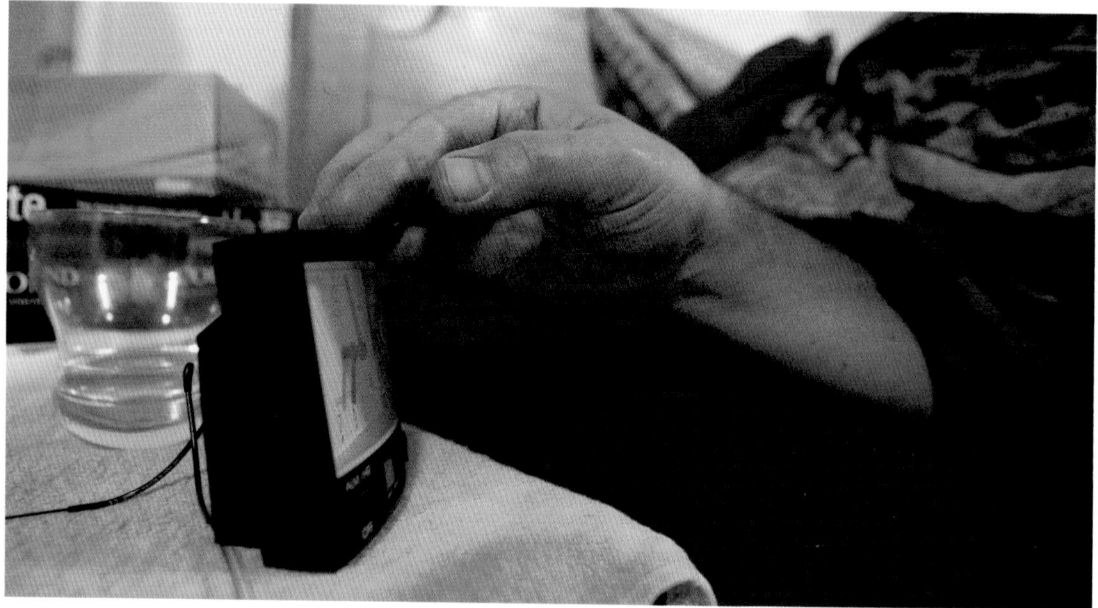

The timekeeping is checked by another clock that knows what time of year it is and can therefore predict when there will be daylight. There is a direct connection between the eyes and the SCN, so the SCN knows when it gets light and dark. These clocks can be fooled. Experiments have shown that people isolated underground in darkness start off with a twenty-four-hour schedule but gradually drift into longer and longer days – twenty-five hours, then twenty-six, and so on – whereas people (and birds) kept in closed rooms with artificial lights that are switched on, say, every twenty-two hours, modify their daily cycles to match the light, provided the changes are not made too quickly. When the basic daily rhythm of light and dark is upset rapidly, you get jet lag, which for me at least seems to be worse flying east than west.

Finally, there is a third group of cells that tell you fairly precisely what time of day it is, which is why many people can decide to wake at, say, 5 a.m. and succeed in doing it.

I have done experiments to find out how much sleep I actually need each night, and after setting the alarm clock a little earlier each evening, I found that I could just survive with six hours. I then slept soundly and felt OK in the morning. However, if I was disturbed in the night, I became tired and irritable, and I certainly could not survive on sleeping for only five and three-quarter hours. In general, I would prefer to have seven hours' sleep each night, and it seems that most people need eight. What is more, we lead such pressured lives that most people are sleep-deprived, which makes them less effective than they could be in their daily tasks and more likely to be uncomfortable and irritable.

ALARM CLOCKS

When I have to sleep away from home, I take an alarm clock, which wakes me with a crescendo of beeping that claws into my dreams. My favourite alarm clock is radio-controlled and therefore accurate to the second – the minute hand clicks over to the twelve as the sixth Greenwich 'pip' sounds. The clock is powered by batteries, regulated by the oscillations of a quartz crystal and checked twice a day by its internal radio receiver, which picks up a signal from Rugby, and so keeps in touch with Greenwich Mean Time, or British Summer Time. It even changes automatically from one to the other when the official change is made.

In the good old days, when time did not seem so important, people in the country were woken by the cock crowing at sunrise. Unfortunately, some cocks crow at two or three in the morning, but I guess that you get used to this and ignore the first few wake-up calls. Recently, on the west coast of Mexico, I heard cocks crowing with enthusiasm in the afternoon; perhaps they are specially trained to put an end to the Mexican siesta. In the early days of the Victorian cotton mills in Lancashire, one or two men from each weaving shed were required to get up long before dawn and go round banging on the bedroom windows of all their colleagues to get them to the mill on time.

Those who wanted to know the time in those good old days used sundials. All you have to do is poke a stick into the ground and lay pebbles where the end of the sun's shadow falls. The shadow is shortest at midday, so if you place pebbles at roughly equal intervals between there and the first shadow of the day – at sunrise – and between the midday pebble and the last shadow of the day – at sunset – you will have a reasonable way of reckoning the passage of time.

Sundials: on the embankment near Tower Bridge in London (above) and at the astronomical observatory in Beijing (below).

Sundials have become immensely more complicated than this. The stick, known as a gnomon, is often aligned with the earth's axis and so is upright only at the North and South Poles. Thus, in Britain, the gnomon is normally set at around 55° to the horizontal, since that corresponds to our latitude. There are sundials carved on stone walls and highly mathematical ones and one in India so big that you can read the time to the nearest minute. Sundials could never be much more precise than this, since the sun is not a point of light; it is wide enough to make shadows of all tall objects quite blurred.

But however big, mathematical and intricate the sundial, it's not much use in cloudy weather, and useless at night. What's more, sundials don't make good alarm clocks – although there were a few months in my life when the sun woke me most effectively. In my late teens I spent a year teaching at a boarding school in India. During the early summer months I slept outside on the veranda in a wood-and-rope bed under a mosquito net. The sun came up at exactly 6 a.m. When the Indian sun hits your white mosquito net, it feels as though a searchlight has been pointed straight at your face and you wake up instantly. But even that would not be much use if you wanted to wake at any time other than 6 a.m.

Intrigued by the idea of automatic waking devices, I decided to investigate their history. The first real alarm clocks we know about were made by an ingenious Greek inventor called Ktesibios, who lived in Alexandria, in Egypt, in the third century BC, and invented the keyboard and the water organ, among other things. The keyboard was – and still is – a mechanical device to control a mechanism using your fingers – I am using one to control my computer, and so write this sentence. The water organ, also called the *hydraulis*, was played with a supply of air kept under pressure by water so that it could play continuously, just as the bagpipes can produce continuous notes because the air supply is kept under pressure in a flexible bag. Water organs later became popular with the Romans, who used them for incidental music during their gladiatorial games.

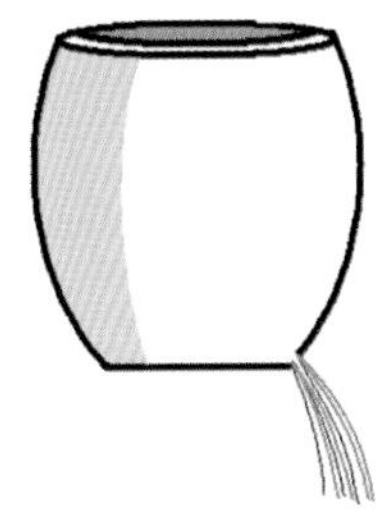

The klepshydra, or water thief: in court, the defendant could speak until all the water had run out.

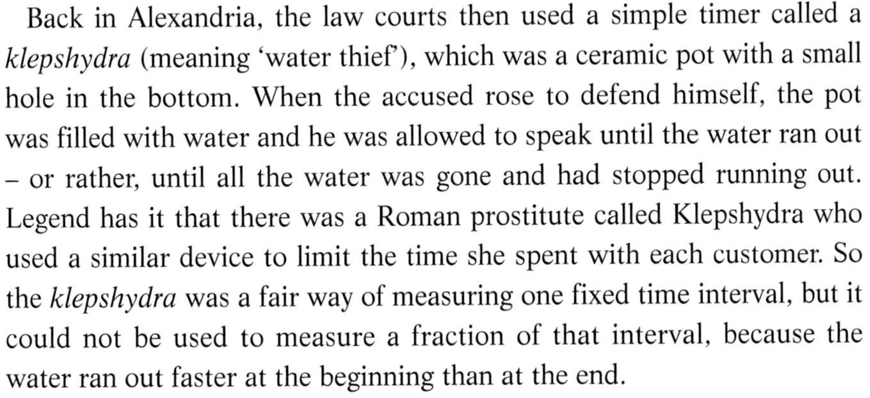

Back in Alexandria, the law courts then used a simple timer called a *klepshydra* (meaning 'water thief'), which was a ceramic pot with a small hole in the bottom. When the accused rose to defend himself, the pot was filled with water and he was allowed to speak until the water ran out – or rather, until all the water was gone and had stopped running out. Legend has it that there was a Roman prostitute called Klepshydra who used a similar device to limit the time she spent with each customer. So the *klepshydra* was a fair way of measuring one fixed time interval, but it could not be used to measure a fraction of that interval, because the water ran out faster at the beginning than at the end.

Ktesibios cunningly got round this problem by arranging for the jar to be continually topped up to the brim, so that the water ran out at a constant rate. He collected this water in a cylindrical jar, in which he also put a float, and the height of the float indicated the time. He might, for instance, start the water flowing in the morning and arrange that the float climbed at a steady hand's breadth every hour; then he could put an hour marker every hand's breadth up the cylinder. Thus he could measure the hours of the day, and because the float was moving upwards, he could set it up to trigger all sorts of devices to make a sound – pebbles that fell on to bells, whistles, even bird calls. These were the first mechanical alarm clocks.

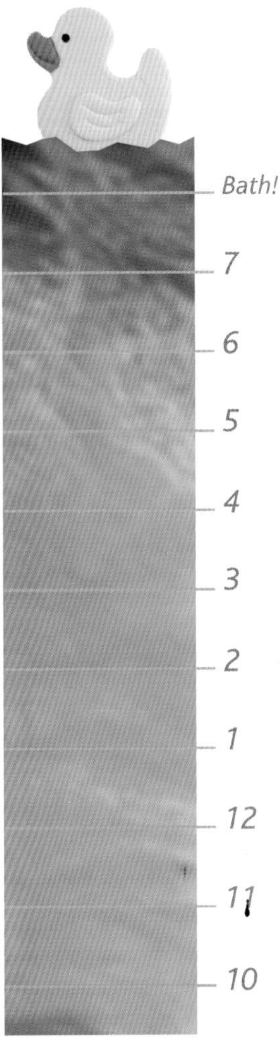

For the television series *Local Heroes*, I had the chance to demonstrate a modern version of one of Ktesibios's clocks in the Greek amphitheatre at Alexandria. What I perhaps should have foreseen was that the white marble of the amphitheatre was exceedingly bright in the midday sun, and we were practically frying as we filmed the gadget. But it still worked a treat.

For the early Greek timekeepers, there was a slight problem in that Greek hours had varying lengths. Each day, the Greeks divided the time between sunrise and sunset into twelve equal hours. The result was that the hours were much longer in summer than in winter, and any one clock was accurate on only two days a year. This may have been why Ktesibios marked his time only in hours; he did not bother with minutes.

The ghati – an early Indian timekeeper. Those in the temples sank after exactly 24 minutes.

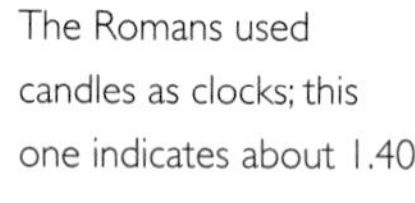

Water clocks like this were the most precise timekeepers for hundreds of years. The Indians came up with their own version, called the *ghati*, which was a bit like a *klepshydra* in reverse. They took half a coconut shell – or a little bowl – drilled a small hole in the middle and then placed it in a bucket of water, like a toy boat. Gradually, the water trickled in through the hole, and as the shell flooded and sank, the priest who looked after it sounded a gong, picked the shell out and floated it once more. The size of the hole was carefully chosen so that the *ghati* sank after exactly twenty-four minutes, or sixty times in one day; the Indian day had sixty 'hours' of twenty-four 'minutes' each. The priest had to act as a manual alarm system, but according to legend, one priest became so in tune with the rhythm that he could strike the gong at precisely the right moment without even looking at the *ghati*.

The Romans used candles as clocks; this one indicates about 1.40.

There is also a sad story attached to the *ghati*. A beautiful princess called Lilavati had been told by the astrologers that she could never be married, but her father, a mathematician, worked out that there was one perfectly propitious moment for her to marry. He made her an elegant *ghati* that was carefully designed so that the moment it sank would be the right instant for her marriage. As the water trickled slowly in, the girl, naturally nervous with her future at stake, leaned trembling over the instrument, and a pearl from her wedding dress dropped into the shell and blocked the hole. So the propitious moment passed unnoticed, and the princess remained single for ever.

The Romans used candles for clocks. In principle, this was simple because the candles could be made with the hours marked on them, but in practice, they cannot have been accurate, since candles are liable to burn at a variable rate – for example, they burn much more quickly in a draught. They were perhaps kept in still rooms, or in glass chimneys, but even so I expect the Roman hours were a bit irregular. To construct a candle alarm clock, however, is simple. Suspend a weight above a gong or bell on a piece of string which is tied around the candle – or passes through it – at the hour marked VI. Then at six o'clock the string will burn through and the weight will fall on the gong and wake you up.

Around the fifteenth century, sailors began to use hourglasses, or sand timers; they used a thirty-minute hourglass to mark the interval between bells, and a thirty-second one to measure the speed of the boat through the water. A lump of wood was thrown over the stern on the end of a long piece of rope with knots tied in it every 15.5 metres (50 feet). The number of knots that went over the rail in thirty seconds was the speed of the ship in knots, or nautical miles per hour (1 knot is 1.15 miles per hour). On land, the Puritans delighted in preaching sermons of two hours, timed by hourglasses. Allegedly, Queen Victoria introduced an eighteen-minute hourglass in self-defence.

In mediaeval Europe, the water clocks and hourglasses were eventually replaced by machines, which had three parts – a power source, an oscillator to provide accurate time and a counting mechanism to count the oscillations. For most early clocks, the power source was a heavy weight on the end of a string, wound on an axle. As this weight descended, the string pulled the axle round, which drove the cogwheels of the clock.

The first important oscillator was the verge-and-foliot escapement mechanism, but this was not very precise. The next big step forward was inspired by the great Italian scientist Galileo Galilei while he was a student in Pisa. He was in the cathedral, listening to a boring sermon, and he noticed the great bronze lamp swinging on its chain from the high ceiling. He used his pulse to time the swing of the lamp and discovered that it always took the same number of beats, whether it was swinging a long way, in the draught from the door, or just a hand's breadth. He realised that a swinging pendulum is an exceptionally regular timekeeper, and by doing some experiments,

Galileo was inspired by the swinging lamp in Pisa Cathedral to investigate the pendulum.

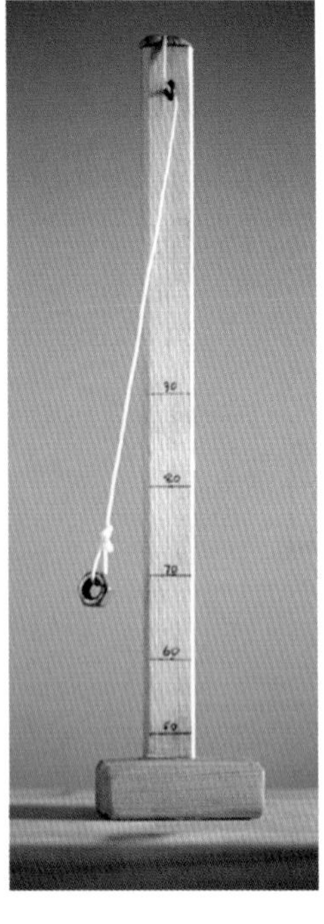

While still a medical student, Galileo invented a simple pendulum timer to measure a patient's pulse rate.

he discovered that the time of the swing depends neither on the weight nor on the width of the swing, but only on the length of the pendulum. I have repeated these experiments in the same cathedral and confirmed Galileo's findings, but sadly, although there is still a great bronze lamp suspended on a long chain, it is now fastened up so that it cannot swing in the draught.

While he was still a medical student, Galileo invented a little pendulum timer for measuring the pulse rate of patients. All he had to do was adjust the length of the string until the pendulum swung in time with the pulse, and then he could read off the pulse rate from a scale written on the shaft of the pendulum. The university authorities stole this invention and gave him no credit. However, Galileo also realised that a pendulum could form the timekeeper for an accurate clock and got around to designing one, but did not finish making it before he died. His idea was taken up by a clever Dutch scientist, Christiaan Huygens (which the Dutch pronounce 'Howgens'), who built the first pendulum clock in 1656.

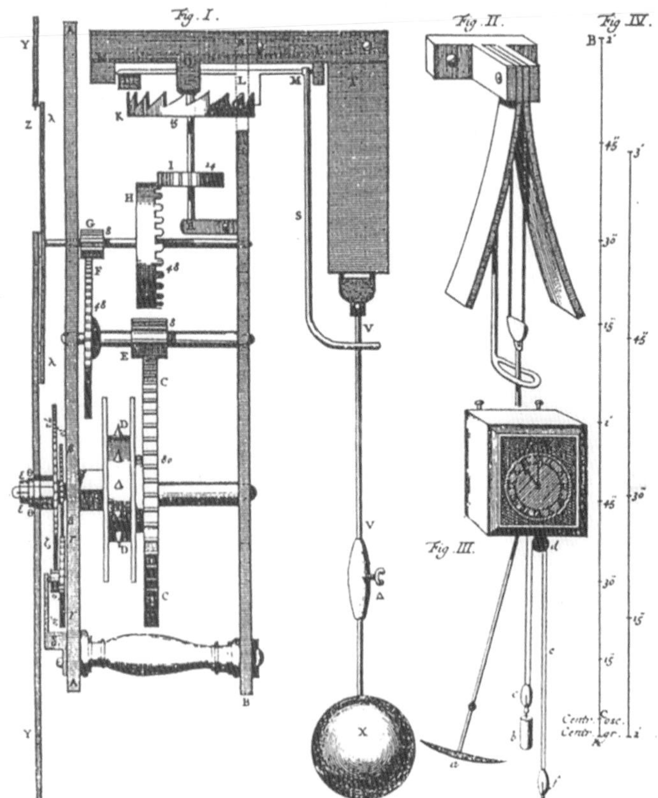

The first pendulum clocks were made by the Dutch scientist Christiaan Huygens.

The next step forward in clock-making was taken by a genius called Robert Hooke, who was born on the Isle of Wight in 1735, worked with Robert Boyle, the father of chemistry, and was for many years Curator of Experiments at the Royal Society. In order to couple the swing of the pendulum to the counting mechanism of the rest of the clock, Robert Hooke invented the anchor escapement, so called because the top part of it looks like an anchor, and because on each half-swing of the pendulum it allows one tooth of the next cogwheel to escape underneath it. The clever thing about this type of escapement is that on each swing the falling weight gives the pendulum a little push and therefore keeps it swinging, as long as the weight does not reach the end of its travel. This was what I used on the clock I once built using Meccano, which kept good time for a minute or two. Hooke's anchor escapement was immensely effective, which is why, with minor modifications, it has lasted for hundreds of years.

The anchor escapement: one tooth of the cogwheel escapes each time the pendulum swings from one side to the other.

Among the modifications was the use of compensating pendulums. A pendulum made with a simple brass rod will run more slowly in summer than in winter. This is because brass expands when it is heated, and so in summer the pendulum will be slightly longer than in winter and the clock will run slow. In principle, this could be adjusted every day by changing the length of the pendulum, but a more cunning solution was to make the pendulum compensate for itself.

Brass expands relatively more than iron when the temperature goes up; so by making the pendulum of several rods, iron rods going down and brass rods going up, clockmakers were able to make pendulums of which the length remained the same whatever the temperature. Another trick was to mount at the bottom of the pendulum a vertical tube containing mercury. As the temperature rises the pendulum lengthens, but the mercury also expands, effectively raising the centre of gravity, and so compensating.

Village carpenter John Harrison, who built the first accurate marine chronometer.

Robert Hooke, father of the anchor escapement, also invented, probably in 1675, a new precise oscillator – the balance wheel and spring, or watch spring – which is used in place of the pendulum in smaller clocks and watches. The advantage of the watch spring over the pendulum is that it does not depend on gravity pulling in a constant direction, so it can be used in watches that are carried about and at sea, where the rolling of the ship would make a pendulum clock useless.

John Harrison went further in his search for the perfect marine chronometer. In 1707, as the result of a navigational error, Admiral of the Fleet Sir Clowdisley Shovell managed to sail the fleet at full tilt into the Scilly Isles, which cost the navy four ships and the lives of 2,000 sailors. Queen Anne then offered a prize of £20,000 for anyone who could solve the problem of longitude at sea, and Harrison, a village carpenter and enthusiastic clockmaker, spent more than fifty years perfecting a sequence of extraordinary and innovative clocks. After building three huge clocks, H1, H2 and H3, he wound up with H4, a pocket watch of elegance and precision, which eventually won him the prize in 1772. All Harrison's clocks can be seen at the National Maritime Museum in Greenwich.

H4, John Harrison's masterpiece, or as he put it, his 'watch, or timekeeper for longitude'.

Harrison was immensely proud of his achievements, and wrote, 'Verily I may make bold to say, there is neither any mechanical or mathematical thing in the world that is more beautiful, or curious in texture, than this my watch, or timekeeper for longitude, and I thank Almighty God that I have lived so long as to in some measure complete it.' A replica of H4 went round the world with Captain James Cook on his third great voyage. Cook didn't survive the trip – he was murdered on a beach in Hawaii in 1779 – but the watch returned less than one minute wrong after three years at sea.

Pendulum clocks are still with us today – in the form of grandfather clocks, such as the fine old beast I have in my sitting room, and the big city clocks including the one in the Palace of Westminster, the great bell of which is Big Ben. I once climbed the tower and stood beside Big Ben while it struck ten. The noise was terrific, and I could feel the structure of the tower vibrating as the hammer hit the bell, although curiously the first of the quarter bells was more unpleasant to my ears.

When it was built in the 1850s, the Westminster clock was required to remain accurate to within one second a day, and surprisingly, it has achieved this. Adjustments need to be made from time to time, and the clock-minder does this by adjusting the number of pennies on a small tray on the end of the pendulum. Adding a penny slows the pendulum very slightly, not because the pendulum bob is heavier, but because the penny on the end slightly lowers the centre of gravity of the pendulum and thus makes it effectively a tiny bit longer.

Big Ben, the great bell of the Westminster clock.

The Westminster clock is extremely reliable and has stopped only a few times in its 150-year history, but it stopped on 27 May 2005 – the day after I visited. I think this was just coincidence.

And what of my radio-controlled, analogue/digital, quartz-regulated alarm clock? Well, it's a direct descendant of this long line of timepieces. Like all mechanical clocks, it has three main components: a power source, an oscillator and a mechanism to count the oscillations. In this case, the power source is two AA batteries, which drive an electric motor, rather than a weight on a string turning an axle. The oscillator is a quartz crystal, which is coated with two metal electrodes and is driven as a capacitor by the battery; it vibrates exactly 33,000 times every second. The counter is electronic; it counts these vibrations, and every 33,000 it moves the display on by one second.

The radio control is an added refinement: the quartz oscillator might vibrate only 32,999 times per second or, in other words, be only 99.997 per cent precise. If that were so, my clock would lose about fifteen minutes a year. When I was a lad, this would have seemed extraordinarily accurate, but now some scientists demand perfection, and we are happy

to make use of it. So the clock has a tiny radio receiver and picks up the Greenwich Time radio signal, transmitted from the National Physical Laboratory (NPL) in Teddington through the BT transmitter in Rugby. We can hear this signal on the radio as six 'pips', of which the last and longest is on the hour. In practice, we hear the 'negative' version: the tone is transmitted continuously, so that you can pick it up at any time of day, and the 'pips' are actually silences or holes in the transmission.

Using this signal, the radio receiver automatically keeps the clock exactly in time, changes over to and from Summer Time and even compensates for the occasional wobble in the earth's rotation, for which the perfectionists have to insert or take away a second every now and then. We used to define time by the duration of the earth's rotation – a second was one 86,400th of a day – but today's clocks are far more regular than the earth. The second is now defined as 'the duration of 9,192,631,770 periods of the radiation corresponding to the transition between the two hyperfine levels of the ground state of the caesium 133 atom'. The National Physical Laboratory gets its time from a caesium atomic clock, the most accurate clock in the world, said not to lose as much as a second in 400,000 years. So I can be fairly sure I shall be woken at the right time, at least for the rest of my life.

THE BEDSIDE LIGHT

Artificial light was one of the greatest inventions of all time. We humans are incompetent in the dark, and without any source of light are almost out of action for a large part of the day. Having a fire for cooking helps slightly, but the invention of oil lamps and candles must have transformed countless lives. Today, these naked flames are no longer important in rich countries, but they are still vital in the developing world.

When the alarm clock crashes into my sleep, I roll over, wallop it into silence and switch on the bedside light, which sounds simple and yet even 120 years ago it was not possible. Domestic electric lighting really began with the invention of the light bulb by Joseph Wilson Swan in 1879, and took many years to become widespread.

In 1799 the Italian scientist Alessandro Volta had found that he could generate an electrical voltage by putting together two different metals, such as silver and zinc. The news spread rapidly and soon British scientists were making huge batteries from which they could draw a steady current. This was a major advance. Scientists had been interested in static electricity and lightning for centuries and had written books about it in the eighteenth century, but now for the first time they had access to a steady supply that they could control.

At the Royal Institution in London, Cornishman Humphry Davy made an arc lamp by shorting a battery through two carbon electrodes and then pulling them slightly apart. To keep the arc going is quite difficult, as the electrodes have to be kept perhaps a millimetre apart, while the arc itself is so violent that it continually crumbles pieces off the tips of the electrodes, thus increasing the gap. Clockwork mechanisms were devised to maintain the gap, but this was tricky. The arc lamp was used, for example, in Dungeness Lighthouse in 1862, but it needed a huge current, was expensive to operate and was far too bright for domestic lighting. The Victorians needed a way to, as they put it, 'subdivide the electric light'.

People had known since the early 1800s that a metal wire heats up when a current is passed through it – this is called incandescence – but the wire usually melts or burns out before it gets hot enough to glow brightly. To make a successful incandescent lamp, the challenge was to find a material for the filament that would not melt at 2,000°C, when it would be white hot, and to keep it away from the air so that it could not burn.

The problem seemed to be soluble in principle in the 1880s when new vacuum pumps promised to remove all the air from a glass bulb, and two rival inventors took on the challenge. Joseph Wilson Swan was a quiet chemist in Gateshead, in the north of England, who worked alone at home in his spare time. I have visited his house to see where he studied the problem. Thomas Alva Edison was a professional inventor with an industrial research laboratory and a large team of scientists in Menlo Park, in upstate New York. He took out more than a thousand patents in his lifetime.

Joseph Wilson Swan, inventor of the first effective light bulb.

Both realised that the critical difficulty was finding the right material for the filament.

Copper, for example, melts at 1,083°C, when it only glows bright yellow, and in any case will burn in air before it reaches that temperature. Swan tried various other metals, including platinum-iridium, which was almost good enough but not quite. Today we use tungsten, but in the 1880s tungsten was hard to obtain and impossible to work. Eventually, both Swan and Edison came to the conclusion that carbon was the best available material. Carbon does not melt below 3,500°C, although it burns easily if there is any trace of oxygen about. The question was how to make strong filaments of carbon, and how to keep the oxygen away.

Swan took narrow strips of paper, some spread with syrup or treacle, and carbonised them by heating them in an oven without oxygen – a bit like burning toast. Some of these filaments glowed quite brightly when mounted in the bulbs, but they were too weak to survive for more than a minute or two. Meanwhile, Edison was convinced that all he needed was the right sort of bamboo: carbonise the fibres and he would have perfect filaments. He sent men to South America, Japan and China in search of bamboo – some died in the attempt – but still his filaments burned out.

Swan eventually concluded that natural fibres made from paper or bamboo would never work. Every one would have some small irregularity, which when the carbonised filament was heated, would become a hot spot and would eventually yield. So he set out to make artificial filaments. He dissolved blotting paper in a solution of zinc

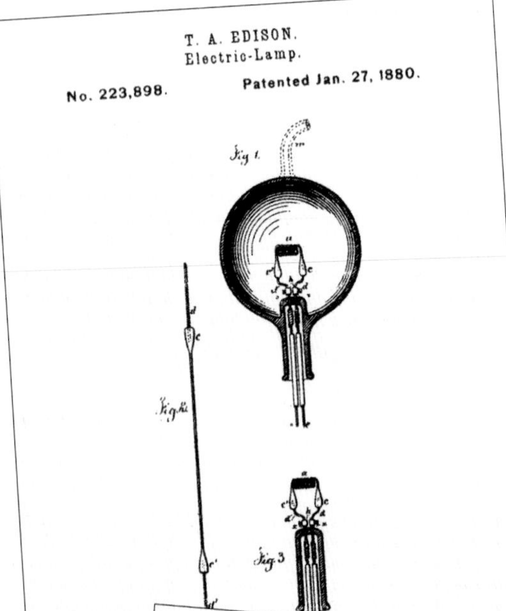

Edison's ineffective patent for the electric lamp.

UNITED STATES PATENT OFFICE.

THOMAS A. EDISON, OF MENLO PARK, NEW JERSEY

ELECTRIC LAMP.

SPECIFICATION forming part of Letters Patent No. 223,898, dated January 27, 1880.

Application filed November 4, 1879.

To all whom it may concern:

Be it known that I, THOMAS ALVA EDISON, of Menlo Park, in the State of New Jersey, United States of America, have invented an Improvement in Electric Lamps, and in the method of manufacturing the same, (Case No. 186,) of which the following is a specification.

The object of this invention is to produce electric lamps giving light by incand-
to which lamps ...

dimensions and good conductors, and a glass globe cannot be kept tight at the place where the wires pass in and are cemented; hence the carbon is consumed, because there must be almost a perfect vacuum to render the carbon stable, especially when such carbon is small in mass and high in electrical resistance.

The use of a ...

chloride; then from a syringe he carefully squirted the mixture into alcohol, which precipitated a long string, or filament, of pure cellulose. These were the first ever artificial filaments, and he persuaded his wife to crochet some into doilies or collars. These still exist and are on display in the Newcastle Discovery Museum.

Swan had foreseen that because they were artificial, his new filaments should have no irregularities and therefore should not develop hot spots. Even so, he found that he had to pump all the air from the bulb, heat the filament slowly and keep pumping to remove the air that had been adsorbed on the filament. Once he had done that, he found his filaments lasted for many hours, and he demonstrated the first useful light bulbs to 700 people at the Newcastle Literary and Philosophical Society in February 1879. Later that year he lit his own house with electricity.

Meanwhile, the ambitious Edison had plans to electrify the whole of New York with large generators and miles of wiring, and he patented the carbon-filament light bulb in October 1879, even though he had failed to make one that worked. When he heard in 1881 that Swan was manufacturing and selling light bulbs, he threatened to sue, but Swan said that there was no point in taking out a patent, since everyone knew the principle of incandescence and in any case Edison's patent was invalid because Swan had been making carbon-filament bulbs for years. Eventually, Edison gave up and they joined forces to form the Edison & Swan United Electric Light Company.

The coiled coil of an incandescent light bulb filament.

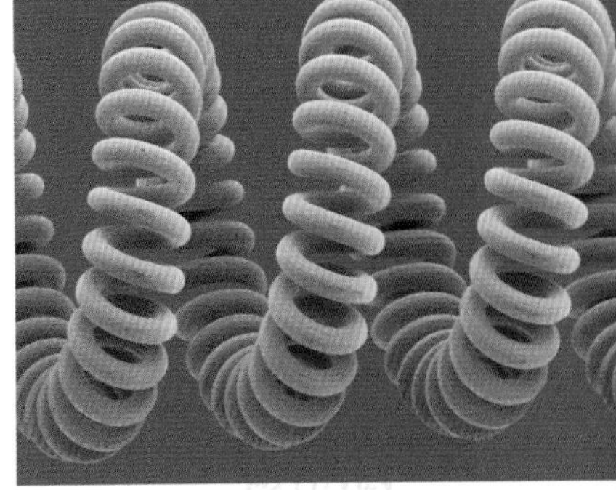

Today we use filaments made of tungsten metal, twisted into coiled coils – look at one with a magnifying glass. This allows a great length to be packed in and holds the separate coils close together, which helps to keep the temperature high and the bulb bright.

Fluorescent light bulbs work not by incandescence but by gas discharge. When the voltage is high enough, electricity will discharge through air, but if the air is dry, it takes 30,000 volts to make a spark a centimetre long. In other words, air is a good insulator. Other gases, however, will allow electricity to discharge at much lower voltages, especially when the pressure is low enough. Low-energy bulbs are filled with such gases. The electrical discharge produces ultraviolet light, and the white phosphorescent material on the inside of the tube absorbs this and glows with a light that is close to white.

Old-fashioned fluorescent tubes, or strip lights, have to be more than a metre long because they are essentially radio valves and they operate at the mains frequency of 50 hertz – mains electricity in the UK is alternating current, which goes backwards and forwards fifty times per second. Modern low-energy light bulbs, or CFLs (compact fluorescent lamps), contain a little oscillator that runs at about 1,000 hertz, which means that the tubes can be much shorter.

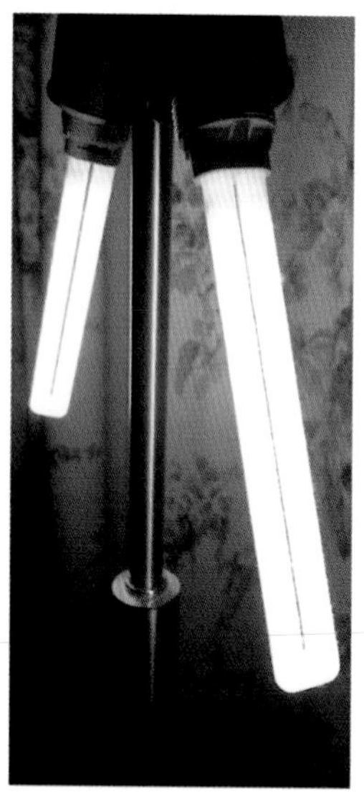

Compact fluorescent lamps, or CFLs (above), are much smaller than old-fashioned strip lights (above right).

Fluorescent lamps are slow to come on because they have a starter-circuit system that takes about a second to kick in. Once the discharge begins, however, the light is often whiter and more like sunlight than is possible with an incandescent bulb. More importantly, these lamps are much more efficient: for a given amount of incoming electric power, discharge bulbs produce three or four times as much light as tungsten-filament bulbs, which waste around 99 per cent of the energy as heat; that is why they get so hot. What's more, low-energy bulbs last about five times as long as old-fashioned incandescent bulbs.

One man who would not have appreciated a bedside light was Christopher Pinchbeck Junior, who, in 1768, took out a patent for a quite bizarre little device. He developed the 'Nocturnal Remembrancer, by which, though the mechanism is very simple, every person of genius, business and reflection might secure all their night thoughts worth preserving, though totally in the dark' – in other words, it would

Never again forget those brilliant ideas in the night, thanks to the 'nocturnal remembrancer' of Christopher Pinchbeck Junior.

help you remember any brilliant ideas you had in the middle of the night.

I made one of these for myself. In practice, it was a glorified notepad, housed in a slim case, which could be made, he said, of 'gold, silver, ivory, tortoise-shell, leather' or any other suitable material. In the front of the case was a letter-box-shaped slot. When you woke with your inspiration, you wrote it down on the pad through the slot, although you had to do this in the dark, which is not easy. Then – and this was the clever part – you could slide the pad up inside the case so that what you had just written was hidden, and if you woke later with another thought, you would not write on top of the first one. If the bedside light had existed, this would not have been a problem and his invention would have been worthless.

Mind you, I don't lie awake worrying about it, since I doubt whether he made much money in any case.

2 Washing and Brushing Up

EXERCISE

Those of a nervous disposition may prefer to skip the next two paragraphs.

When I feel energetic, which I like to think is on most days, but in reality is probably twice a week, I go out early for some exercise. Ideally, I leave the house at about 6 a.m. and return at about 7 a.m. During the summer I usually cycle a fairly standard circuit that includes the Bristol Downs, where there are joggers, dog-walkers and often bird-watchers looking out for the peregrine falcons that nest in the Avon Gorge. In mid-winter it is still dark at 6 a.m., so I take a fifteen-minute ride up to the local gym and spend half an hour on various machines, watching the early news on television.

Reluctantly working up a sweat.

Why do I take exercise? Partly because I enjoy it at the time, especially riding around the Downs in the early-morning quiet, watching the birds and seeing the sun come up; partly because I feel better for several hours afterwards, and my skin is improved, presumably because sweating clears out the pores; and partly because I hope it will help me to lose weight. The good feeling comes from the increased levels of serotonin in the blood; this is like an endorphin, a pleasure drug produced by the body. Besides, exercise is meant to be good for you. It improves blood flow and lowers the blood pressure, thus reducing the risk of stroke and heart disease In addition, it increases bone density, improves blood-sugar levels, and burns off fat. I like to take some exercise every day, but doctors recommend a minimum of thirty minutes of vigorous exercise, such as brisk walking, at least three times a week.

When I feel less energetic, or the rain is bucketing down, I go straight to my computer and get on with some work, usually first tackling any new email that has come in during the night. I generally feel great early in the morning, but I must remember not to make any important decisions too early: recent research at the University of Colorado has shown that people perform extremely badly when they first wake up. This grogginess, or sleep inertia as Professor Kenneth Wright calls it, is severe for the first few minutes, but the effects can be detected for up to two hours.

Whether or not I have been out panting and sweating, my first duty of the day is to make the tea at a quarter past seven for my partner, Sue Blackmore, and myself. I don't mind much whether or not I have tea. I drink it when I have made a pot, but when I am on my own, I often don't bother. Sue, however, in common with many others, needs her cup of tea to start the day. Other people rely on coffee to wake them up. Coffee contains caffeine, while tea contains theobromine; both are stimulants and provide a kick-start to the nervous system, although they seem to have little effect on mine.

When I take up the tea things on their tray, I cannot help remembering my favourite clue from *The Times* crossword: 'The first swallow? (5,7,3)' (hint: it's on the tray). I am not a crossword fanatic, although I have once completed *The Times* on my own, but I saw my dad finish it in the ten-minute interval between the innings at Lord's Cricket Ground. On another occasion my stepmother was reading out the clues, and she said 'The next word is three blanks I, four blanks I,' whereupon he interrupted 'The answer's "Garibaldi". What's the question?'

Early morning tea.

THE SHOWER

Many years ago I spent a year in India at a school where the shower was simply a cold tap overhead. There was no 'hot' water, but in the middle of the summer, in April, the water from the 'cold' tap was so hot I could not stand underneath it.

My shower now has hot and cold supplies, with a mixer valve to adjust the temperature and a shower head on the end to make a fine spray. It also has a pump to push the water out in a satisfying gush. This is a luxury, but also a necessity, since the shower head is within 20 centimetres of the roof and there is nowhere to place a water tank higher than this, so my shower cannot be powered by gravity, as most showers are. I have a shower every morning; it helps me wake up, and afterwards I feel clean and fresh and ready for the day. Usually, I take my shower first thing, but when I go out for some early exercise, I shower afterwards to remove all the sweat and grime.

Why are showers so satisfying? First, the water hits me – especially my face – in small drops and at a high velocity, so that I get a feeling of being thoroughly cleansed. As a former chemist, I know that repeated rinsing with small volumes of water (or other solvent) is much more efficient for cleaning than a single slosh with a large volume. Second, the amount of water I use in, say, five minutes is much less than I would use in a bath deep enough to cover my body, which means that I am saving both water and heat. Third, the water hits me at a constant temperature, whereas in a bath the temperature is always dropping. Fourth, the soap and dirt are continually washed away by the flow of water in the shower, whereas in a bath I have to sit in the mess.

My shower has a glass door, but often in hotels I have to use plastic shower curtains, which tend to attack me and wrap themselves round my body. This is a result of the Bernoulli effect, first described by

Power showers are wonderful, but you may be attacked by the plastic shower curtain.

mathematician Daniel Bernoulli in around 1730; he said that when a gas flows quickly, its pressure is lowered. The rush of water in the shower drags down with it a rush of air, and because this air is flowing rapidly, its pressure is lower than that of the air in the rest of the room. The excess air pressure outside the shower curtain pushes it in towards the shower, and I usually get in the way, only to be wrapped in wet plastic. The remedy is either to leave a gap at the end of the curtain to let air rush in there, or make sure your face and upper body intercept most of the water; then there will be much less rush of air, and the curtain will not billow.

One of the most charming showers I have enjoyed was out in the woods. Hung from a tree branch was a plastic bucket with holes drilled in the bottom. I heated a large kettle of water on the campfire, tipped it into the bucket and stood under it surrounded by trees and birdsong, with a long view across the valley. The cleansing may have been less than perfect, but the experience was delightful.

One intriguing fact about showers is that the water is warmer in the middle than at the outside. I have measured two showers and found that the middle water is about 2 to 3°C warmer than the edge water. Why should this be so?

Hot water always tends to evaporate if it is in contact with cool air, and a hot shower will allow much more evaporation than a hot bath. This is partly because the water in the shower comes in thousands of small drops, which together have a much greater surface area than the bath and so evaporate much more quickly, and partly because each drop falls through the air for nearly half a second. The steam that is formed by the evaporation condenses on cold surfaces, which is why after a shower the walls, mirrors and other cold surfaces in the bathroom get covered in tiny droplets of water – condensation.

As it comes out of the shower head, the water in my shower (set to maximum temperature) is at 50°C. It hits me at about 49°C, and by the time it gets to the shower tray, the temperature has fallen to 46 or 45°C. But this is not surprising, since warm water will always cool down by evaporation. To turn water into steam requires energy, and this energy can come only from the water itself. Therefore, as some of the water in each little drop evaporates, it extracts heat energy from the rest and so cools the drop.

This, however, does not explain why the water should be warmer in the middle. This curious effect was first pointed out to me by a meteorologist as we stood in the showers after a game of squash. His explanation, which I suspect is correct, since meteorologists spend lots of time thinking about temperature and evaporation, was as follows. In the middle of the spray, there are many drops, all evaporating, and the air between them must be almost saturated with water vapour. This means that little more water can evaporate, because there is no 'room' for it in the air – or to be more precise, almost as much steam is condensing from the air on to the drops as water is evaporating from them. Therefore the drops will cool only slightly. On the outside of the spray, however, there is plenty of cool, dry air, which allows the drops to evaporate more quickly and cool faster, and so the shower will always be cooler on the outside than in the middle.

The worst shower I ever had was in a cheap hotel north of Rome. We arrived at around 5.50 p.m. after a long and tiring drive, only to be told by the receptionist that there was no hot water after 6 p.m. I ran up to my room, flung open my suitcase, grabbed my sponge bag, leaped into the shower, turned it on full and smothered my head with shampoo. At that moment the shower head came off, the water spurted up on to the ceiling, over the top of the shower curtain, and for fifteen deadly seconds went straight into my suitcase. Then, with a clunk, the water stopped altogether, leaving me dry and covered in sticky shampoo, and my suitcase full of wet clothes…

SOAP, GEL AND SHAMPOO

Getting clean needs more than water alone. Human skin is usually both damp and greasy, because of the oily sweat that comes out through the pores. Dust and dirt in the air stick to this damp grease, and in order to get rid of it, you need to wash away the grease. Grease is not soluble in water, so water alone will not wash it off.

The long molecules of grease are made mainly of hydrogen and carbon, which makes them hydrophobic – they do not mix with water. What you need is some sort of detergent or 'surfactant' – a compound with molecules that have both hydrophobic ends, to mix with the grease, and hydrophilic ends, to mix with the water. The best-known such material is soap, first developed by Arab scientists in the Middle Ages by boiling oil with alkali. (*Al-qali* is an Arabic word that originally meant the ashes of the plant saltwort.) The alkali breaks the hydrophobic oil molecules and adds a hydrophilic chunk to the broken ends. As a result, soap will form a stable emulsion with oil or grease and still dissolve in water, so it will wash the grease off your skin, carrying the dirt with it.

To make solid soap, the Arabs used caustic soda as their alkali, and modern hard soaps, as well as such detergents as washing-up liquid, shower gel and shampoo are made of sodium stearate and similar sodium compounds. All these products have additives to provide colour, scent and texture, which make them more attractive.

The surfactants have to be just powerful enough to dissolve the grease without damaging the skin. If they are too powerful, they react with the lipids in the skin; this breaks down the water barrier so that the skin dries out and cracks or chaps. This is why your hands can get dry and sore if you wash too many clothes or dishes by hand.

I learned recently that food can have a direct effect on your voice, which might be important to me when I sing in the shower. Eat a mouthful of chocolate and your mouth feels all claggy and slimy. This is the fat coating your gums and the inside of your cheeks. Try to sing through a mouth lined with chocolate and it sounds terrible. Wash your mouth out with water and nothing much happens, because the fat is not soluble in water, unless you put a squirt of lemon juice in the water. Then the result is instant: your mouth is clear in a flash and feels sparkling clean. This is rather like the effect of those surfactants in shampoo, except that the lemon juice is not just mixing with the fat but reacting with it to make it soluble. That is why vinegar is good for cleaning glass. Both lemon juice and vinegar are mildly acidic, and the acid reacts with the fat to make products that are soluble in water.

Of all the various cleansers available, I am a believer in soap. When I get into my morning shower, the first thing I do is wash my face with soap, but some people, especially women, wash their faces in a different way from their bodies. Many women use only cold water on their faces, without any soap or face wash, which is probably a good idea for skin that is on the dry side of normal, while for greasy skin face wash may be better. Face wipes, I am told, cleanse and do other good things in one go.

One skin problem that causes much distress is acne, which appears as whiteheads, blackheads and red pimples or zits. Caused neither by eating greasy food or chocolate, nor by poor hygiene, but by an imbalance of hormones, acne is often a nightmare for teenagers and for women at times of menstruation or pregnancy, because increased levels of hormones in the bloodstream increase the amount of oil formed in the follicles under the skin. This extra oil can mix with dead skin cells and clog the pores. Dirt and bacteria will then collect underneath to make all those ugly lumps. Whiteheads are essentially closed – you can see the white dead skin cells – while blackheads are open to the air, which oxidises the melanin in the skin cells and makes them look black. Squeezing the lumps may be satisfying, but does not remove the underlying cause and can lead to further infection and even scarring.

You may be able to prevent acne, or you may be able to treat it. The simplest preventive measures are to wash the face regularly – perhaps twice a day – with mild soap that does not dry. Avoid oily or greasy make-up and hair products, and take off all make-up at night. If this does not work, you can buy creams or lotions over the counter at the chemist. These contain such chemicals as sulphur, benzoyl peroxide or salicylic acid, which kill the bacteria and dry up the oil. If even these do not work, your doctor may be able to prescribe antibiotics or other medicines, but make sure you get your own individual prescription, since some of these drugs can be dangerous if taken by the wrong person. Contraceptive pills are helpful for some women, but make the problem worse in others.

Rather than soap, many people like using shower gel or body wash, which contain many of the same ingredients as soap, but are formulated to remain liquid and thick, so that a measurable quantity can be squeezed out of the container. With soap, on the other hand, you simply have to rub it in your hands for a time and you cannot tell how much you are using. Shampoos contain similar surfactants and fragrances, but they also contain ingredients specifically designed for hair. In 1986, for example, manufacturers introduced the 2-in-1: shampoo and conditioner combined. Now, almost all shampoos include conditioner, the function of which is to keep your hair strong, smooth and flexible.

Different people have different ideas about what 'clean' means: some like to be left with an oily sheen; others prefer a squeaky cleanness. Washing in hard water produces squeaky cleanness much more easily than washing in soft water. To cope with these various possibilities Western consumers want a range of products.

Everyone has to buy soap or shower gel and especially shampoo on a regular basis, and so there is tremendous pressure on manufacturers to make a product that is seen to be better than its rivals. That is why there are so many brands and why each brand offers several variants. These used to be products for greasy hair, dry hair, frequent use and so on, but now they often focus on the outcome – perfect curls, extra volume (which means less conditioner), lustrous shine, etc. I recently went into a large chemist's shop in the town centre and counted no fewer than 257 allegedly different types of shampoo – and that ludicrous number does not include the proprietary brands sold only through hairdressers.

Why does my local chemist stock 257 allegedly different varieties of shampoo?

So, is it all nonsense and hype? Not quite. The manufacturers do try to give you the best, and what you want. In the average shampoo there may well be twenty ingredients, each doing a specific job, and all carefully mixed in a gloop that flows from the bottle but does not instantly run off your hand and down the drain.

Electron micrograph of a single human hair sprayed with dry shampoo (false colours).

WASHING YOUR HAIR

I thought washing and drying my hair was a simple task, but when I consulted the experts, I found I was doing it all wrong. Here is some scientific and expert advice on how to wash and dry your hair, especially if it is long:

1 Treat your hair gently, as you would your face or a cashmere sweater.

2 Brush or comb it first.

3 Wash it long, starting at the top and working down the length of the hair. Never pile it up and rub it with shampoo on top of your head. You probably have 150,000 individual hairs; tangle them and you are in trouble.

4 Use lots of conditioner, especially at the ends. You can't use too much, unless you have 'spaghetti hair' that always hangs straight and you long to give it more body.

5 Rinse it with lots of water.

DRYING YOUR HAIR

Wind hair round curlers while drying it and it should stay curly until it gets damp again.

After my shower, I usually dry my hair with a towel, using the 'friction' method I learned from a teacher at my prep school, when he was supervising our communal baths – in a large room with six baths in it. I hold one end of the towel in each hand, flop it across my head and see-saw the towel from side to side. I should point out that this method is heavily frowned upon by the experts, because it could damage the hair, but my hair is short and tough, and does not fall out in handfuls.

The women in my house use a hairdryer, which is basically an electric fan with a tube that blows out a jet of air. It also has a heating element, so that the jet of air can be heated. Water from wet hair will slowly evaporate into the atmosphere. This process is encouraged if a jet of air blows away the water vapour, and happens even more quickly when the air is hot, as the hot air warms the water on your hair and warm water evaporates faster than cold water. Indeed, if you use cold air, your head will cool down, because the evaporating water extracts heat in order to turn into vapour. This is why you feel cold when you get out of the shower, the bath or swimming pool. A hot-air dryer keeps your hair warm, which is more comfortable as well as being more effective. Use that hot air with care, however…

DRYING YOUR HAIR

How to dry long hair (advice from the experts):

1 Squeeze it dry. Do not wring or you will tangle it and break it; hair is weakest when wet. Pat it dry with a warm towel.
2 Use the hairdryer slowly, on 'warm' rather than 'hot'. Blasting wet hair with a hot hairdryer will make the hair shrink and crack.
3 Finally, comb or brush your hair gently, starting with the ends and working slowly back up.

By drying your hair carefully, you can change its form – make it straighter or curlier, at least for a time. Wind your hair round curlers before you dry it and the hair will remain curly until it gets damp again. Your hair is also likely to get slightly more curly if you let it dry naturally in the air, especially by the sea, when it may be salty. This effect is caused by hydrogen bonding. Hydrogen bonds are weak attractions between different molecules, rather like magnetic attractions. They hold molecules of water together and give water its surface tension and its high boiling point.

Hydrogen bonds between different molecules in your hair hold it in one form – which for most people is straight. When the hair gets wet, the water molecules float the hair molecules apart and temporarily break these hydrogen bonds. Wind the hair round curlers while you dry it, and when the water evaporates, the hydrogen bonds will form in new places between the molecules in your hair and so keep it curly.

I never try to make my hair curly, but I have used much the same trick in my green woodwork. Every summer I try to get away to the woods to make a piece of furniture and learn more about working with green wood (see page 62). Sometimes this involves bending pieces of wood. Two years ago I made a kitchen chair with a back made from bent sweet chestnut. Last year I made a tray with folding handles made of bent ash (see photo on page 25). First, I kept the wood in steam for an hour or so, and then, with help, quickly bent it round a former to get it to the shape I wanted. The steam has the same effect on the wood as the water on your hair – it breaks the hydrogen bonds between separate molecules in the wood and so allows the

Electron micrograph of a single human hair, more flexible and easier to cut when wet.

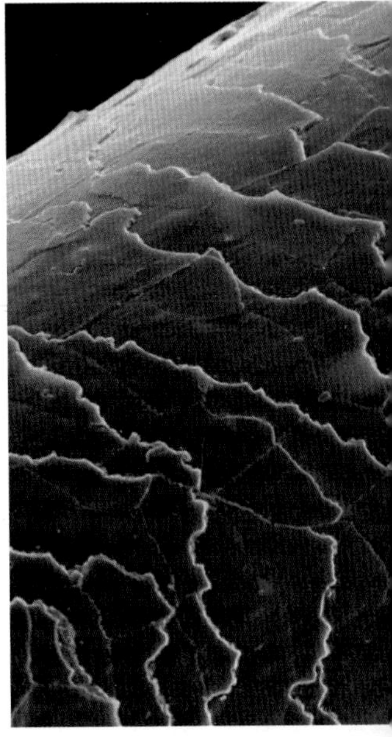

fibres to slide past one another. After the wood has been bent, the steam escapes and the fibres bond together in the new shape. The wood still has some tendency to spring back, but remains strong in its bent shape.

MIRRORS

One question I am often asked is, why is it that when you look at yourself in a mirror you are reversed left to right but not upside down? The simplest answer I have come across is that you are not really reversed left to right, but rather front to back.

Look in a mirror from say half a metre away and you see a reflection of yourself apparently the same distance behind the mirror. Let's call this reflection Mirror Person, or MP. Now imagine that MP is actually looking in the opposite direction, so that you are looking at the back of MP's head. Then you would see the back of MP's left ear on the left side, and the back of the right ear on the right. Now imagine that MP's head is pushed through itself: the ears don't move, but the nose is pushed through from the far side to the near side and so on. You are now looking at MP's face, but what appears to be MP's right ear is actually the left ear reversed front to back. Touch your own left ear and you will see MP touching the ear on the left. In fact, the whole image in the mirror has

Mirrors reverse images from front to back. These pictures reproduce the effect I saw at Frankfurt airport.

been reversed front to back, but we perceive that it has switched from left to right (see also page 143).

Recently, I was presented with a vivid example of this at Frankfurt Airport, where the name stands in letters a metre high in front of a wall of reflective glass. Even when you see it in the mirror, the writing is still the right way round, because you see the reflection of the backs of the letters, and they are reflected back to front.

The letters are not really reversed from left to right. Instead the reflection of the green backs of the letters reads the same as the front.

Curved mirrors present a slightly different puzzle. Look at your reflection in the back of a shiny spoon – a convex mirror – and MP, now small and distorted, is reversed back to front as usual but still the right way up.

In the concave hollow of the spoon, however, MP is distorted and reversed both upside down and from front to back. Do try these experiments for yourself, preferably in a magnifying mirror, which is much easier to work with than a spoon. First, if you can get close enough – easy with a magnifying mirror but difficult with a spoon – you will see a magnified image: MP is larger than life and reversed front to back, as in the flat mirror. Move slowly away from the mirror and you will reach the point of focus, where everything goes rather blurry, and your eye may fill the entire mirror. This is the maximum magnification the mirror can provide. Your eye fills it because the curvature of the mirror is such that whichever point of the mirror you look at, your line of sight meets the mirror at a right angle, and so you see your eye reflected.

Move further away and the image becomes clearer. Now you can see MP reversed upside down and front to back. Touch your left ear and upside-down MP will touch the ear on the right. This is not quite the same as being rotated by 180°, because if you hold up some writing, you will see that even when you rotate it 180°, it is still back to front in the mirror. Look into the left side of a distant concave mirror and you see a reflection of the right-hand side of the world; look into the top and you see a reflection of the bottom.

CLEANING YOUR TEETH

Why should we need to clean our teeth at all? Animals don't bother; I have never seen my cats worrying about tooth decay. The culprit may have been the invention of farming, and the fact that people have therefore been eating bread, porridge and similar products for some thousands of years. Perhaps these foods have encouraged tooth-attacking bacteria to thrive in our mouths and we have not yet evolved enough to escape their attention.

In 1977 the American physicist Richard Feynman painted a delightful word picture of tooth-brushing. Imagine the world, he said, as seen from outer space. Curving from one Pole towards the other is a line of sunrise: to the west of the line, the earth is dark; to the east, it is light. All along this line people are getting up and brushing their teeth – millions of people, all brushing away. This line of tooth-brushers sweeps around the globe, moving along the equator at 1,000 miles per hour, and it never stops.

Feynman went on to say that there was no evidence that all this brushing had any effect in preventing cavities, but some dentists disagree. Teeth are useful for talking and eating, and dentists urge us to take good care of them, which, they say, means brushing them at least twice a day and flossing or using inter-dental brushes at least once.

Pig-bristle toothbrushes were all the rage in the late nineteenth century.

The first toothbrushes were probably twigs, and Mesopotamians were chewing sticks 5,000 years ago. Some people still use twigs, and the modern toothpick is essentially a modified twig, but the notion of using a brush, with bristles embedded in the end of a stick, seems to have occurred first in China. The earliest known toothbrushes, in around AD 1000, were of horsehair in an ivory handle, but later ones were of pigs' bristles in bone. The idea spread gradually, and in 1649 Sir Ralph Verney was asked to bring back to England from Paris some of those 'little brushes for making cleane of the teeth, most covered with silver and some few with gold and silver twiste'. By 1780 William Addis was mass-producing them in England; the bristles, from Siberian and Chinese pigs, were embedded in holes drilled in the cow-bone handles. Americans don't seem to have picked up the toothbrush until the mid-1800s.

In around 1940 pig bristles began to be superseded by nylon filaments, which are cheaper, more consistent and more hygienic, since bacteria do not cling so easily to the synthetic material. Soon after nylon filaments appeared, we were offered soft, medium and hard versions, and then the nylon handles began to be made in various colours. As with shampoo (see page 28), there is tremendous selection pressure to make a better toothbrush, and so the supermarket shelves are littered with perhaps fifty kinds, of various shapes, sizes, makes and models, all competing in a fierce evolutionary contest. There is even a new 'Twist 'n' Brush' with a rotatable head designed to get at the back of the front teeth. The very profiles of the bristles are now fiercely patented by the manufacturers. Dentists in developed countries tend to recommend brushes with medium or preferably soft filaments, and small or streamlined heads with many bristle tufts and bristles round the end. The problem is getting into the awkward corners, and what really matters is the dexterity of the brusher.

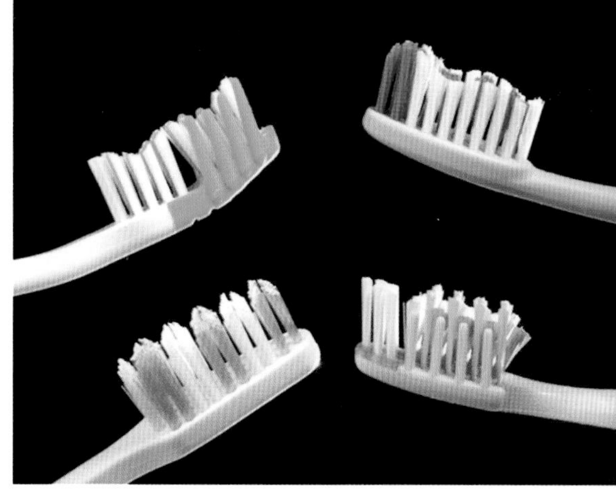

Even the profiles of the bristles are patented today, but we have long had considerable variety in our toothbrushes.

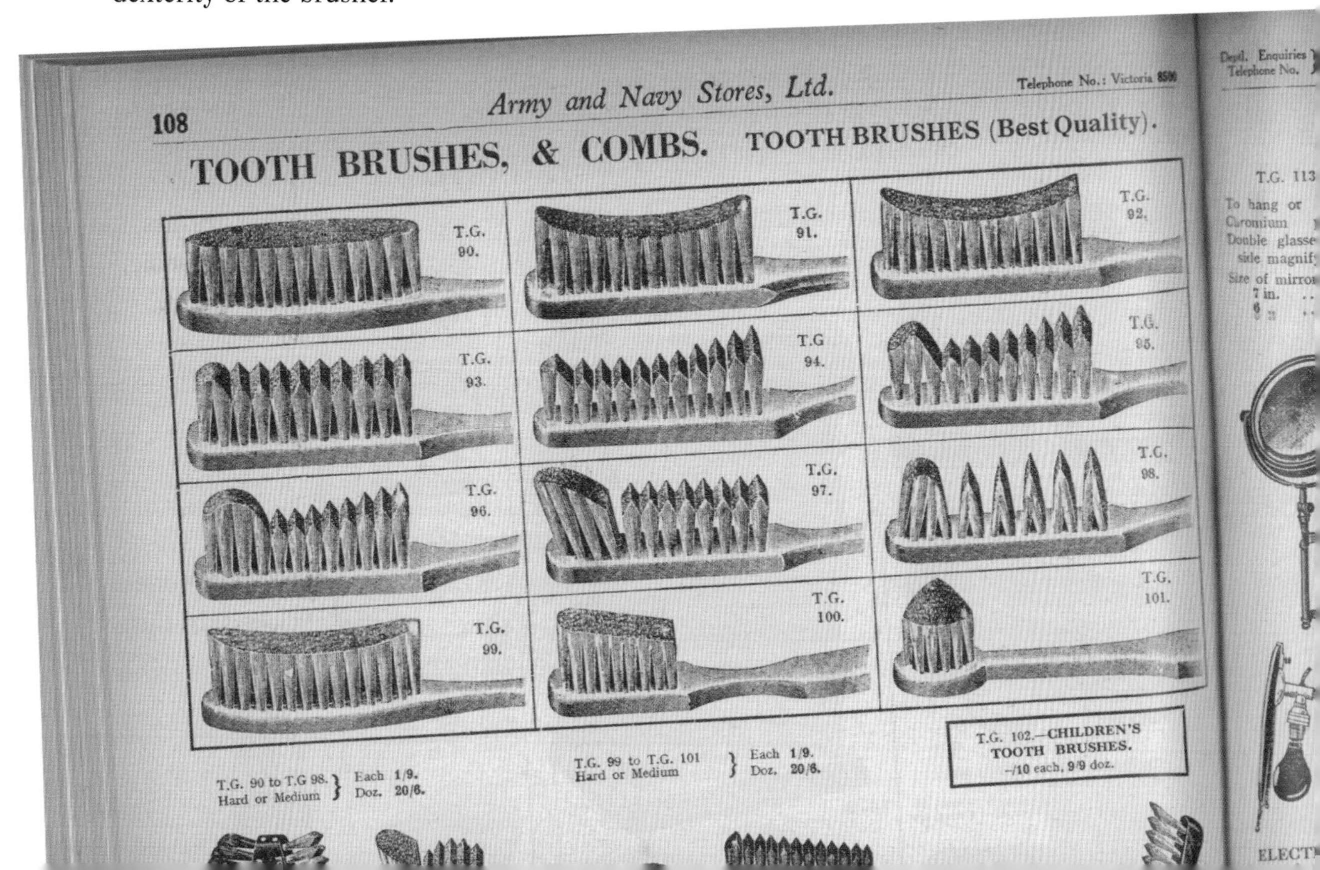

108 *Army and Navy Stores, Ltd.* Telephone No.: Victoria 8500

TOOTH BRUSHES, & COMBS. **TOOTH BRUSHES (Best Quality).**

T.G. 90. T.G. 91. T.G. 92.

T.G. 93. T.G 94. T.G. 95.

T.G. 96. T.G. 97. T.G. 98.

T.G. 99. T.G. 100. T.G. 101.

T.G. 90 to T.G 98. Hard or Medium } Each 1/9. Doz. 20/6.

T.G. 99 to T.G. 101 Hard or Medium } Each 1/9. Doz. 20/6.

T.G. 102.—CHILDREN'S TOOTH BRUSHES. -/10 each, 9/9 doz.

BRUSHING YOUR TEETH

What is the best way to brush your teeth? These instructions are from the website of the American Dental Association (see www.ada.org for details):

1 Place your toothbrush at a 45-degree angle against the gums.
2 Move the brush back and forth gently in short (tooth-wide) strokes.
3 Brush the outer tooth surfaces, the inner tooth surfaces and the chewing surfaces of the teeth.
4 Use the 'toe' of the brush to clean the inside surfaces of the front teeth, using a gentle up-and-down stroke.
5 Brush your tongue to remove bacteria and freshen your breath.

Electric toothbrushes first appeared in Switzerland in 1939 and reached the United States in 1960. The modern ones have small, circular bristled heads that rotate or oscillate in two or three dimensions. Some are battery-powered. I use one with sealed rechargeable batteries inside. The recharging mechanism is cunning: it works by electrical induction, so the brush itself need not have any electrical contact on the outside. No one likes the idea of getting a shock from a toothbrush.

Dentists say that electric brushing may be more effective than manual. In particular, the expensive machines with complex movements are more efficient at cleaning teeth. I find mine makes brushing easier, especially behind the lower front teeth, and it has a timer that tells me, by means of coded vibrations, when I have been brushing for two minutes, which is definitely longer than my normal manual brushing time. However, experts tell me that most of the plaque is removed in the first thirty

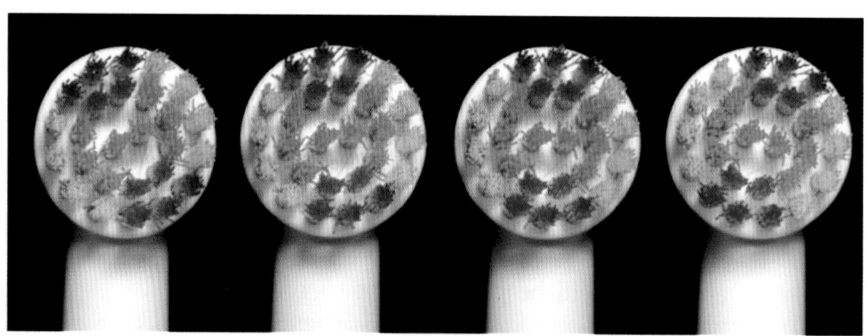

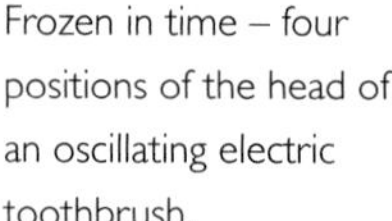

Frozen in time – four positions of the head of an oscillating electric toothbrush.

seconds of brushing and it always returns in the same places, because we fall into set patterns when we brush our teeth. Perhaps we should occasionally brush one another's teeth, and then the patterns would keep changing; but I don't really think that is a practical solution.

I have always thought that the main point of brushing my teeth was to prevent tooth decay, but I was corrected by a professor of oral hygiene. In fact, brushing has relatively little effect in preventing tooth decay. The main point of brushing is to prevent gingivitis, or inflammation of the gums.

Each time you eat, particles of food lodge between your teeth and along the edges of your gums. This food encourages the growth of bacteria, of which there are always millions in your mouth. The bacteria consume the food, especially any sugar, and turn it into acid, which is what causes the trouble. The bacteria live in a sticky layer of biofilm called plaque, which you can often feel with your tongue as a roughness on your teeth. Regular brushing prevents plaque from building up. If you don't clear away the plaque, it will harden into tartar, which can't be shifted with a toothbrush but needs a dentist or dental hygienist with a scraper or scaling instrument – my dentist uses an ultrasonic air-scaler, which uses a jet of compressed air to vibrate a small metal point to and fro at high speed, while a fine mist of water cools the point.

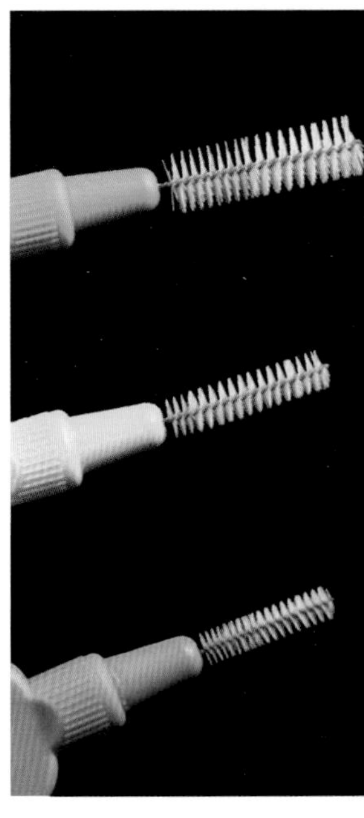

Interdental brushes range in width from about 2 mm to 5 mm; these are the thicker ones...

There is a further problem in the narrow gaps between your teeth and in tiny fissures in the teeth themselves: underneath the plaque, the acid made by the bacteria attacks the enamel, which is the tough outside coating of your teeth. If this process is allowed to continue, the acid will eat right through the enamel and expose the sensitive dentine inside, forming first a white spot and then a painful cavity.

Because plaque tends to build up not on the shiny front surface of the teeth but in the narrowest gaps, it is hard to remove it all with a toothbrush. You can use dental floss, which is thin nylon string, to help clear the plaque from these gaps. Although it was not actually patented until 1874, dental floss was strongly advocated in 1819 in a book by an American dentist called Levi Spear Parmly. He wrote of a waxed silken thread 'to be passed through the interstices of the teeth, between their necks and the arches of the gum [...] With this apparatus thus regularly and daily used, the teeth and gums will be preserved free from disease.' Unfortunately, floss is difficult to use properly, and indeed tricky to use at all, and

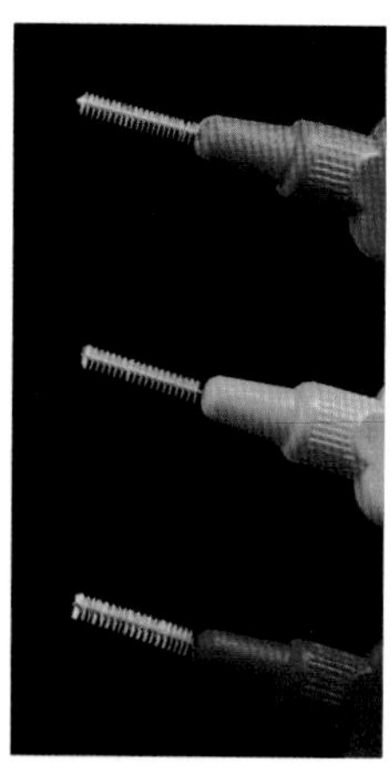

... and these are the thinner ones.

although three million miles of the stuff are sold in the United States every year, in practice only about 15 per cent of the population flosses regularly, or in other words the average American flosses only about once a week.

Floss comes in various thicknesses and shapes: some is in ribbons and some is braided, so that it is thin when stretched but widens out to do a better cleaning job when the tension is reduced. Dentists recommend flossing at least once a day. The idea is to take a 20-centimetre length of floss, wind the ends around the first or second fingers of each hand, and then push it down through each of the gaps between your teeth, one by one. At the bottom, you should slide it under the edge of the gum to lift out any particles of food or traces of plaque. This may be reasonably easy to do in someone else's mouth, where you can see what you are doing, but not in your own, and therein lies the problem.

Because floss is hard to use properly, my dentist has recently advised me to change over to brushing with inter-dental brushes – fine stiff brushes that slide between the teeth and do the same job. This is slightly less fiddly than flossing, and may be more effective at clearing the food from between the teeth, but it cannot clear the edges of the gums as effectively as perfect flossing should do.

Dental floss has other advantages, however; it is immensely strong and can be used as string, clothes line or fishing line, and for all sorts of other jobs, such as sewing on buttons, repairing clothes, wetsuits, slippers, tents, sails and rucksacks, hanging pictures, cutting cakes and cheese and making necklaces. Unstoppable English cyclist Josie Dew says she never leaves home without it; several prisoners have escaped from jail using ropes made from dental floss; and in the north-western United States, prisoner Scott A. Brimble used dental floss and toothpaste to saw through the wire mesh of an Okanogan County Jail exercise yard. He was charged with first-degree escape.

Bad breath, or halitosis, may be caused by failure to brush your teeth, because as well as acid, those bacteria in the plaque can produce sulphur compounds, which smell terrible. However, bad breath is more likely to come from the back of your tongue, since lots of bacteria grow there; the solution is to use a tongue scraper, or to brush your tongue with your toothbrush. Bad breath can also result from various types of infection. Perhaps the worst problem with bad breath is that those who have it cannot smell it, and so do not know there is a problem. the only way to find out is to ask someone else – perhaps your dentist if all other possibilities are too embarrassing.

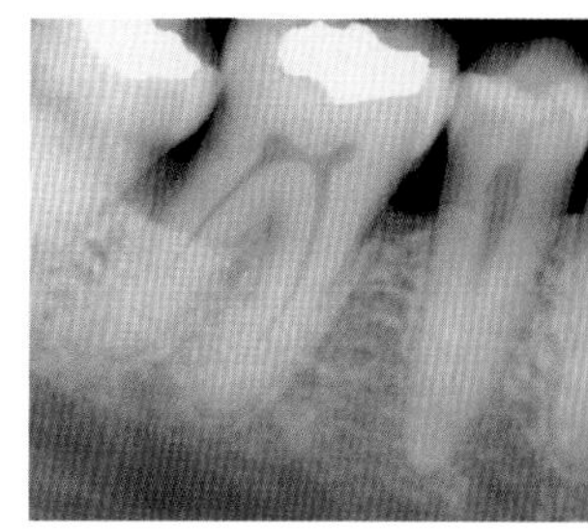

A coloured X-ray of teeth, showing the fillings on the biting surfaces and the roots buried deep in the gums.

Chewing gum is slightly helpful for oral hygiene. The action of chewing stimulates the production of saliva, which helps to flush away the particles of food, and some gums contain mild antibacterial products, as well as breath-fresheners, which make your mouth feel cleaner. In a controlled trial in Budapest, 290 schoolchildren had to chew sugar-free gum for twenty minutes three times every day after meals; one year later they had significantly less tooth decay than their friends who had not chewed gum.

Some people think that swilling with mouthwash will clean their teeth, but most often that's not the case. The mouthwash may kill bacteria on the surface of the plaque, but the ones underneath will carry on growing. A swill of mouthwash may make your mouth taste clean and freshen your breath, but it will not remove the plaque. To get rid of the plaque, you have to scrape it off, and a toothbrush and floss are your best weapons.

How effective is toothpaste? The short answer is not enormously, despite the claims of manufacturers. In a large survey, Scandinavian schoolchildren brushed their teeth regularly with dry toothbrushes and got no cavities at all, suggesting that toothpaste is not necessary. Apart from providing lubrication so that your brush slides easily across your teeth, the main function of toothpaste seems to be to taste good and so to encourage brushing: for children, there are strawberry pastes; for British adults, mint is common, while Europeans seem keener on herbal flavours.

Electron micrograph of crystals of toothpaste abrasive (false colours).

All toothpastes are abrasive: they contain small, hard particles of silica or chalk that slightly help to scrape away the plaque, and in particular help to remove stains. One manufacturer produced paste with no abrasive and received many complaints that it caused staining of the teeth. Tooth powders are abrasive and were common fifty years ago, but seem to have gone out of fashion. Some brushers prefer the feel of gels; they have more water than ordinary paste. The abrasive gels use flakes of silica, which appear transparent, so that the entire gel is translucent. Tartar-control toothpastes contain pyrophosphate, which by some cunning chemistry prevents tartar from settling. There is fierce competition among manufacturers, and many toothpastes offer special ingredients that are supposed to leave your teeth cleaner, whiter and brighter, but it seems that you may as well use salt, powdered chalk, or even soot – anything slightly abrasive will do. However, most toothpastes do contain bactericidal agents, breath-fresheners, desensitising agents, whiteners and, most importantly, fluoride.

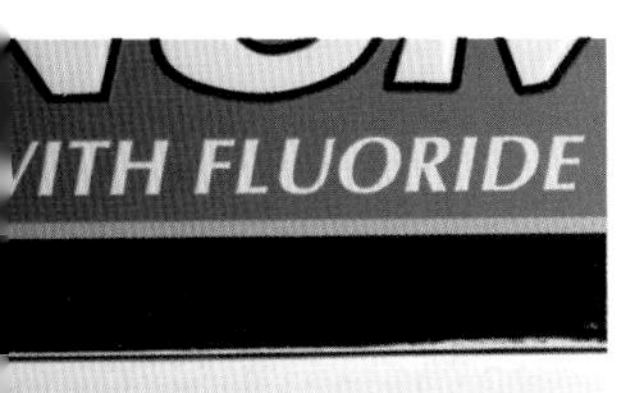

Fluoride is probably the most important ingredient of toothpaste.

Common table salt is sodium chloride, and so is most of the salt in the sea. Fluoride is a close relative of chloride and has been shown to strengthen tooth enamel, which greatly helps to reduce tooth decay. You drink some fluoride whenever you have a cup of tea – it's an entirely natural material. In many parts of the world, there is a minute amount of fluoride in the drinking water and in those places people have better teeth. In some other places, fluoride is added to the water, but it is also a good idea to use a toothpaste that contains fluoride. Too much fluoride can make your teeth brown and brittle, so children should be discouraged from eating toothpaste by the tubeful, but otherwise fluoridation has been, according to the US Surgeon General, 'one of the greatest achievements of public health in the twentieth century'.

Toothpaste tubes developed from those used for oil paints.

Toothpaste tubes originated as containers for artists' oil paints, but in 1892 a dentist called Washington Sheffield started using the same sort of collapsible metal tubes to sell toothpaste. The tubes were originally made of an alloy of lead with a few per cent of antimony, and presumably we all, with our toothpaste, ate some of the lead, which is a cumulative poison.

N° 7402  A.D. 1897

Date of Application, 22nd Mar., 1897
Complete Specification Left, 22nd Dec., 1897—Accepted, 5th Feb., 1898.

PROVISIONAL SPECIFICATION.

Improvements relating to Collapsible Tubes for Containing Liquid or Semi-liquid Materials.

We, John Scott Taylor, of North London Colour Works, Kentish Town, N.W., Technical Chemist, and Winsor & Newton, Limited, Manufacturing Artists' Colourmen, of 38, Rathbone Place, W., do hereby declare the nature of this invention to

Then, when plastics became common, ICI – and no doubt others – spent some research time trying to find a 'deadfold' plastic film – i.e. a material that would stay flat after being squeezed. They tried polythene, polypropylene and polyvinylchloride (PVC), with clever plasticisers, but without success. Today some manufacturers use plastic tubes, which do not stay flat after squeezing; others use a composite of plastic and aluminium, and the resulting deadfold tubes do stay flat.

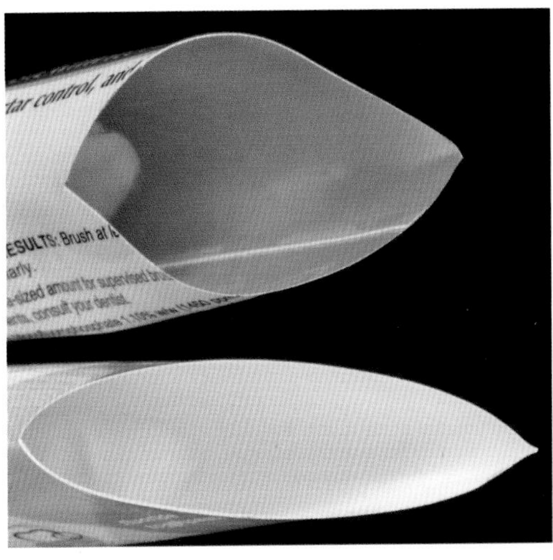

Some toothpaste tubes are made of plastic alone; others have a lining of aluminium.

Squeeze some paste out of a plastic tube and then stop squeezing, and it will spring back to its 'full' shape, sucking some air into the top of the tube. This has two drawbacks. First, air may affect the contents: paint hardens in air, although toothpaste is barely affected. Second, the top of the tube is now full of air, rather than paste; so the next time you use the tube you have to squeeze out the air before you get any toothpaste. A brilliant new idea solved this problem – the big top. Fifty years ago, all tubes had small plastic screw caps, but now most tubes have screw tops 2 or 3 centimetres wide. This means you can stand the tube on its cap and all the toothpaste will slither down to the business end, so the nozzle is always primed with fresh toothpaste. It also makes life easier for the retailer, who can stand dozens of tubes side by side on the shelves, whereas tubes with small caps are difficult to display. All sorts of manufacturers now sell products in tubes rather than bottles, presumably for our convenience – everything from shampoo and conditioner to mustard and tomato ketchup. Meanwhile, air-sensitive products such as paint and glue are still sold in collapsible metal tubes made of aluminium, which is safer and cheaper than lead.

Standing tubes on their big tops allows the contents to slither down to the right end.

MAKE-UP AND MOISTURISER

In July 2005 Prime Minister Tony Blair admitted that every year he spends hundreds of pounds on cosmetics and make-up artists. Quite apart from the fact that this is taxpayers' money, I was surprised to learn how much he cares about his appearance. I use make-up only when it is put on me forcibly before I go into a television studio (see page 157). At home, all I use is an occasional dollop of moisturiser.

Your skin is the largest organ in your body; among its vital jobs are to keep unwanted bacteria out and to keep water in. The human body is about 70 per cent water, and if this drops by only a few per cent, many of your enzymes will stop working and you will die. That is why extensive burns are so dangerous: they destroy large areas of skin. If only to maintain your water level, your skin needs to stay in good fettle.

The topmost layer of skin is a tiling of dead cells, which gradually flake off as more come up from underneath. This is why small dirty marks and scratches disappear in a few days. The top layer normally contains about 10–15 per cent of water, which helps to keep the surface soft and smooth, but this varies with the humidity and temperature. Lie in the bath for an hour and your skin will become wrinkly because it is absorbing water from the bath and swelling up; this does no harm. What matters much more is dehydration. In unusually dry weather, water may evaporate from the surface of your skin faster than it comes up from below, which will make your skin dry up, crack and split. It may become hard and scaly, and pieces may flake off, which is roughly what happens in eczema and psoriasis. Skin may also suffer from prolonged exposure to water, detergents, rubber gloves and other foreign materials.

Electron micrograph of dry skin, showing the tiling of dead cells on the surface.

Putting on moisturiser helps in three different ways to prevent your skin from drying out. Moisturisers are made mainly of water, and adding this water to the skin replaces some of the water that is being lost; but they may also contain humectants, chemicals that help to hold more water in the skin, like chemical sponges. In addition, moisturisers usually contain a waterproofing oil, which sits on the surface of the skin and cuts down the rate of evaporation of water, just as a lid on a plastic coffee cup helps to keep the coffee warm. Some moisturisers contain 'anti-ageing' ingredients, which are generally sun-block materials to prevent damage from the sun's ultraviolet rays.

Using moisturiser therefore does help to keep your skin soft, but do not be fooled by 'miracle' additives. Aloe vera, for example, is a plant from the lily family and is known to have anti-irritant properties, but these will not be effective unless the moisturiser contains perhaps 10 per cent of aloe vera, which would make it outrageously expensive. Manufacturers are keen to sell their products and to persuade customers to pay more, so they will add royal bee jelly, placental extract, turtle grease, shark oil, horse-blood serum, vitamins, seaweed and all sorts of other exotic ingredients to try to convince you. Some of the resulting products are sold for more than £1,000 a jar, and glamorous actresses are said to use nothing else. In my view these exotic concoctions are just another form of 'snake oil' and are no better than simple cheap moisturiser. They will not make your skin softer or sexier. Vitamins, for example, are necessary for the functioning of the body, but cannot be absorbed through the skin; all the moisturiser can do is rehydrate dead cells, although vitamin E may help to reduce the effects of free radicals produced by sun damage.

BODY ODOUR, OR BO

Elizabeth I is alleged to have had a bath every month 'whether she needed it or not'. In Elizabethan times people washed rarely, and body odour must have been the norm; indeed, the entire house must have been disgustingly smelly. Some people carried bouquets of flowers to disguise their own smell. Today we wash ourselves and our clothes more frequently, which probably makes us more sensitive to such unwanted smells as body odour and bad breath.

Unfortunately the worst offenders often don't realise they have BO.

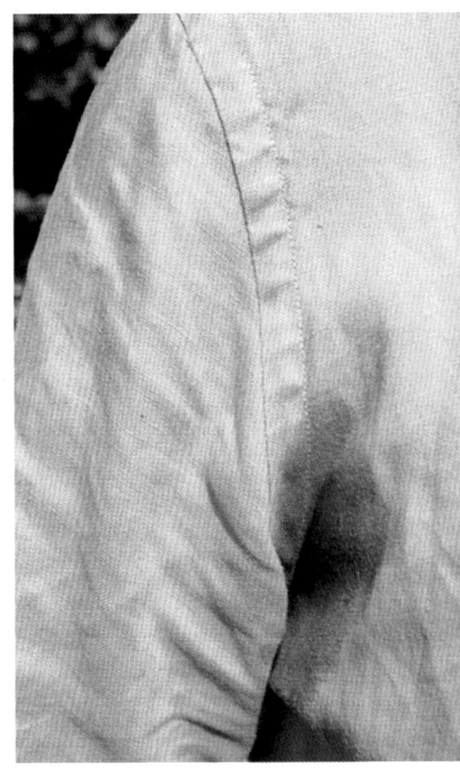

Sweat is made by the sweat glands in the skin and comes up to the surface through pores. When the sweat evaporates, it cools the skin; indeed, sweating is the body's main cooling system. Everybody sweats, although some sweat more than others; I sweat profusely when I take energetic exercise, when I am hot, or when I am concentrating – for example while taking photographs. Not everyone has BO, however, so why do some men smell so foul? How can a single person make it almost impossible for anyone else to share a bus, a lift or a room?

Sweat itself is mostly salty water and does not smell, neither do clean clothes, even when they are soaked with sweat. However, warm sweat does provide a splendid growing medium for bacteria, and they thrive in the warm hollows of the armpits, groin and feet. In these areas, the apocrine glands exude proteins and other chemicals, which bacteria break down to produce ammonia and volatile fatty acids, such as butyric acid and E-3-methyl-2-hexenoic acid. These acids are what smell so foul. Unfortunately, people who produce these smelly acids cease to notice them after a time, and as a result, the people with the worst BO are often unaware that they smell.

The simplest way to get rid of the problem is to wash these areas with soap and warm water, twice a day if necessary. In addition, wearing clean clothes every day is a good idea, for stale sweat on the clothes is almost as bad as stale sweat on the skin. A short-term measure is to apply antiperspirant, which temporarily stops the sweating, or deodorant, which kills the bacteria. The active ingredients in antiperspirants, apart from perfume, are usually aluminium or titanium salts, which have a great affinity for water and soak it up like a chemical mop, with the result that you do not get wet patches in the armpits of your shirt.

Those who cannot control the problem with these simple methods may like to try shaving the armpits, since removing the hair cuts down the surface area for the bacteria to thrive. They may acquire bactericide solution from the chemist, or a solution of aluminium chloride to apply last thing at night; this will block some of the pores and lower the amount of sweat on the following day.

I use neither antiperspirants nor deodorants; I just hope I wash myself and my clothes often enough not to have BO.

SHAVING

The Philishave razor with its floating heads.

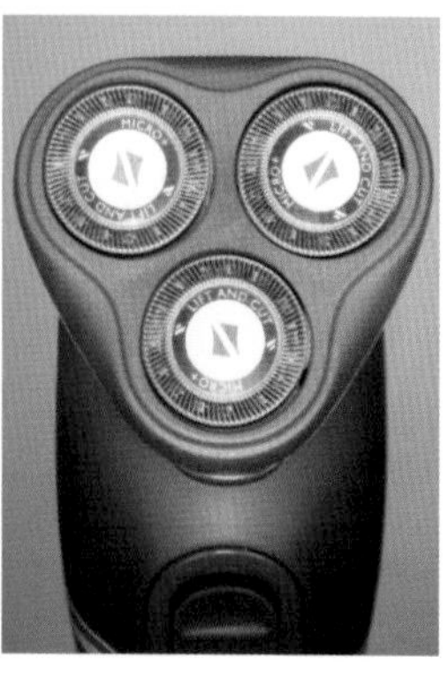

Human hair grows at around 0.000,000,0075 miles per hour, or one-third of a millimetre a day, although the rate varies greatly from one person to another. Most people who want to get rid of it from their faces or elsewhere choose to cut it off. Of the three billion boys and men in the world, roughly 40 per cent do not shave, 15 per cent use electric razors, and 45 per cent use manual razors.

Electric razors are essentially like miniature mowing machines – a thin metal foil perforated with many holes that you hold against your skin, and a rotating or reciprocating blade or blades that move across the other side, to act like multiple pairs of scissors and snip off all the hairs that poke through. A slight problem is that the hairs are cut not at the level of the skin but the other side of the foil, leaving them perhaps 0.03 millimetres long, since the foil is about 0.03 millimetres thick.

There is a huge range of electric razors on the market, with subtle differences in modes of operation. All have a smooth piece of metal foil peppered with perforations about half a millimetre wide, which the hairs of your beard go through. Some shavers have two or three circular shaver heads, with one, two or three rings of cutters underneath the foil. Others have a single head in the shape of a pair of long arches, with the foil the size of a pair of short chunky pencils. Underneath this, the blades go backwards and forwards. In the more expensive models, the entire head oscillates to and fro by about a third of a millimetre, which helps to pull the hairs up and feed them through the perforations in the foil.

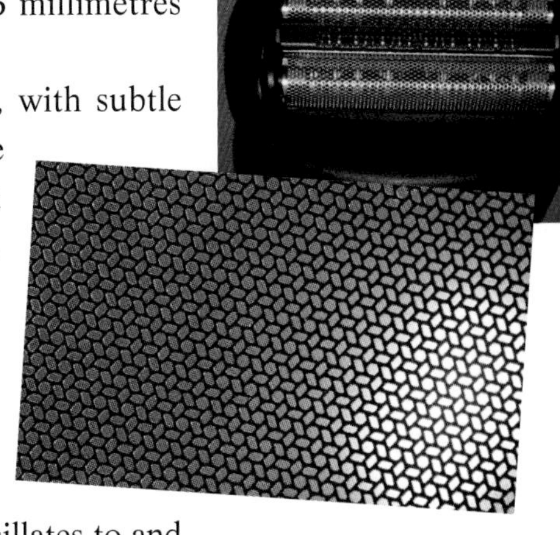

The Braun razor has metal foils so thin that a stack of 30 would make a pile less than one millimetre high.

The great advantage of electric razors is that you simply switch on and shave; you do not have to find all the materials needed for a wet shave. Nor do you need to wash before or afterwards. If you run out of electricity, you can use a battery-powered model or a clockwork razor, which has a cutting mechanism essentially the same as for electric machines, but the power is provided by winding up a spring, so you can use it anywhere.

I have never used an electric razor; when I shaved regularly, I used a brush and shaving cream or oil. I still believe in wet shaving when I am trimming the edges of my beard, which I do about once a week. I use a badger-hair brush and a tube of shaving cream. Although it is rather a palaver, wet shaving leaves me feeling wonderfully fresh and clean and smooth.

The 18 lifters and cutters inside a Philishave head.

Using a razor on dry skin is uncomfortable, because the skin is not slippery, the hairs are tough, and the razor will jump about and remove little bits of skin here and there, making it sore. Wetting the skin lowers the friction and so makes shaving easier and more effective. Wet hair is also softer and easier to cut: ten times as much force is needed to cut through a dry hair as through a wet one. Women who shave their legs and armpits in the bath or the shower tell me that as long as they are thoroughly wet, then ordinary soap is a good lubricant – better than shower gel – and helps the blade slide over the skin.

For men who want to shave their chins, the experts recommend that first you wash with warm water to soften the hairs, and then either gently rub into the skin shaving oil or gel, spray on foam, or apply shaving soap or cream with a soft shaving brush. Oil is the simplest, and particularly good when you are travelling, because you need so little. A teaspoonful will last for weeks, and a small plastic bottle keeps me going for months. Wet the skin and the beard first, then put a couple of drops of oil on the palm of your hand and rub it gently into the skin with your fingertips. Then wet your razor and shave. The oil lubricates the skin, so the blade of the razor slides over it easily and smoothly; the oil may also help to soften the hairs.

Shaving gel, foam and lather all have four functions. First, like oil, they lubricate the skin and enable the blade to slide smoothly over it. Second, they are wet and so keep the hairs damp, soft and easier to cut. Third, the razor sweeps away the stuff in its travels and so provides a useful map of where it has been. Because you can see where the gel, foam or lather has been removed, you know exactly which bits of your skin you have shaved, which is not always easy to tell, especially in the early morning, when your eyes are half shut and the mirror is covered with condensation. Fourth, the cut hairs are swallowed up and swept away by the goo; they do not clog up the razor blade and are easily washed away. In addition, most of these products are gently scented and provide a comforting smell.

Using a traditional badger-hair brush helps to lift the hairs away from the skin.

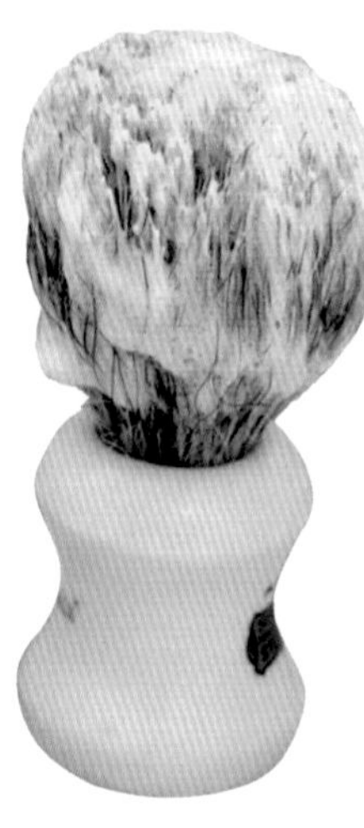

Fifty years ago barbers and most men used shaving brushes and blocks of shaving soap. Rub the wet brush on the top of the block of soap and then apply to the face with small circular movements to work up a rich lather. This feels good and ensures that the skin is well lubricated and that the lather is firm enough to last for several minutes. The brushing has the additional effects of raising many of the hairs that tend to lie along the surface of the skin so that you get a better shave, and also removing dead skin cells that are ready to fall off. Today the block of shaving soap has largely been replaced by shaving cream in a tube, which requires less effort and may be less wasteful, since you can squeeze out the minimum amount you need.

The best shaving brushes are said to be made of badger hair, but this may be just a matter of fashion or folklore. However, long-term wet-shavers care passionately about their brushes. A Canadian friend used to visit England every ten years or so and always made a special pilgrimage to Harrods just to buy himself a new shaving brush – for him, the ultimate luxury.

Shaving gel comes in tubes like toothpaste and is thick and sticky, so that it stays in place when you spread it on your skin with your fingers. It is made partly of water and therefore mixes with the water on your skin, and surrounds the hairs to keep them damp and soft. It also contains some clever polymers that help to lubricate the skin and make shaving more comfortable. When you have finished shaving, you can easily wash off any residual gel with water, which gets rid of all the cut ends at the same time. The advantage is that you do not need a brush – although for me, brushing is part of the pleasure.

Shaving foam is squirted from aerosol cans and is therefore easy to apply – you simply spray it on. Arguably, however, it cannot make such close contact with your skin as lather, which you actively brush in. Foam works in much the same way as lather and gel, but environmentalists might disapprove of the unnecessary use of aerosols.

And finally, the razor: what an elegant piece of high technology, over which so much ingenuity has been expended down the centuries. Razors seem such simple things, and yet because most men use them every day, because they have immediate effects on your comfort and appearance, and because they are short-lived so that you have to buy new razors frequently, there is tremendous selection pressure. In other words, there are vast rewards to be gained by making a razor that is slightly better or cheaper than those of your rivals. This must have been true for centuries.

A 16-cm long sixth-century iron razor found in central Europe.

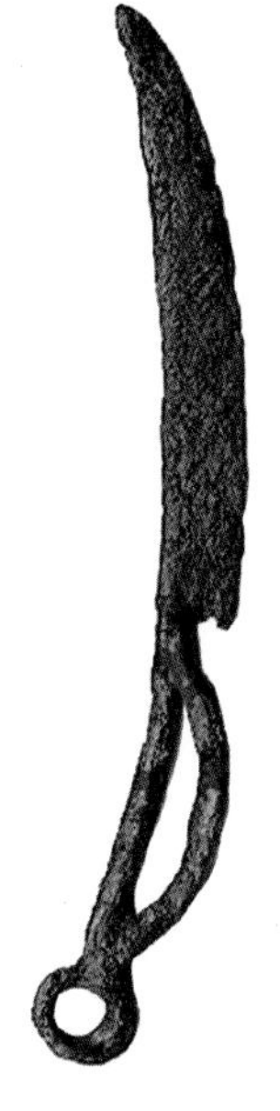

SHAVING IN HISTORY

Men have shaved for thousands of years. Remains of ancient razors have been found in Egypt, India, Greece, Rome and all over Europe. They are made of iron, bronze or copper alloy, and often have broad, flat blades. They were heavy and difficult to handle, and almost certainly men did not shave themselves but went to be shaved by a barber. Barbers' shops became important meeting places and centres for gossip, and some barbers became immensely wealthy. Julius Caesar, in common with other rich men, had his own personal barber and enjoyed a daily trim. Some barbers were less popular, however: the satirist Martial said that one,

Antiochus, was so clumsy that a surgeon hacking through broken bones was more gentle, and that another, Eutralepus, was so slow that a second beard had grown by the time he had cut the first. Barbers' shops were dangerous too – think of the crowds and all that jostling. Before he died incautiously watching Vesuvius erupt in AD 79, the Roman statesman and scholar Pliny the Elder recommended spiders' webs soaked in oil and vinegar to staunch the flow of blood – probably a good ancient equivalent of the modern sticking plaster and styptic pencil.

A 30-cm long Greco-Roman bronze razor from the first century BC: the flat blade was sharpened on both sides, and the wiggly handle shaped to fit the hand.

Some people preferred depilation – to remove the hair, men and especially women rubbed their skins with all sorts of things, including pumice stone (probably the most effective), asses' fat, bats' blood and powdered viper (rather like those magic ingredients in the wrinkle creams, see page 44). There were specialist underarm hair-pluckers in the Roman equivalent of the Yellow Pages, and some cavalrymen shaved their legs, as do some cyclists today, in the hope of going faster.

Alexander the Great, who died in 323 BC, was a trendsetter. Before him, most men wore beards, but he went clean-shaven; other Greeks followed suit, and then so did the Romans, to show how 'Greek' and civilised they were. The long, thick beard then became the trademark of philosophers and other rebels, to demonstrate their intellectual superiority. The Greek doctor Galen, however, in the second century AD, said that beards were good for men because they helped to drain off the ill effects of exhalations of humours that built up inside them – in other words, beards helped to get rid of disease – though he did not explain how.

In sculptures, beards are shown on old men; youths are beardless. This is important in gay Greek culture: a beardless statue, however large and muscular, suggests a desirable boy.

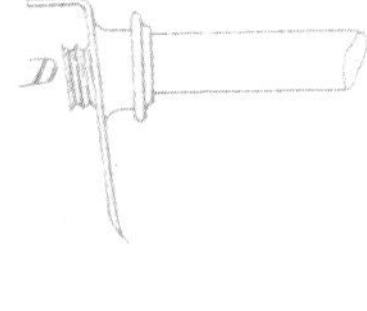

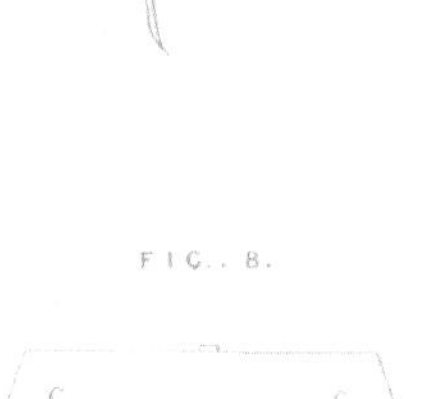

Shaving cannot have changed much for 1,500 years, but the production of better steel in the eighteenth century enabled the manufacture of steel razor blades. Even so, the blades needed frequent sharpening and had to be stropped every day – a strop was a strip of leather against which the razor blade could be polished. Rich Victorians often had a set of seven razors, one for each day of the week. These were 'cut-throat' razors: a curved blade about 10 centimetres long that folded after use into a handle of wood or ivory. However, cut-throat razors were inherently dangerous, and it was only too easy to give yourself a serious injury if your hand slipped. They were also vicious weapons in the hands of thugs or criminals with evil intentions.

In 1770 a French barber, Jean-Jacques Perret, wrote a book called *La Pogonotomie, or The Art of Learning to Shave Oneself*, in which he proposed the use of a safety razor – essentially a cut-throat with a wooden guard to prevent serious injury. He went on to make such razors, which were a moderate success. In 1847 William Samuel Henson, an Englishman who sported a magnificent moustache, patented another safety razor – a cut-throat with a safety comb attached to the blade. Henson had spent the previous few years working in England with John Stringfellow, trying not only to build an aircraft but also to set up an airline, which was greeted with derision. He was doubly unlucky. After Henson left England, Stringfellow managed to get an aircraft to fly, powered by a miniature steam engine. And Henson's safety razor failed to catch on; perhaps it just wasn't good enough. He moved to New York, where a few years later the Kampfe brothers patented yet another safety razor, which again was only moderately successful.

Sam Henson's 1847 patent for a safety razor.

A.D. 1847 . . . . Nº 11

Razors.

HENSON'S SPECIFICATION

TO ALL TO WHOM THESE PRESENTS SHAL
S. HENSON, of No. 27, New City Chambers, in the
greeting.
WHEREAS Her most Excellent Majesty Queen V
Patent under the Great Seal of the United Kingdor
Ireland, bearing date at Westminster, the Seventee
eleventh year of Her reign, did, for Herself, Her h
and grant unto me, the said William S. Henson, I
I, the said William S. Henson, my executors, adı
or to such others as I, the said William S. Henson
trators, and assigns, should at any time agree wi
time to time and at all times during the term of
should and lawfully might make, use, exercise, ar
Wales, and the Town of Berwick-upon-Tweed,
Majesty's Colonies and Plantations abroad, my
IMPROVEMENTS IN THE CONSTRUCTION OF RAZORS FOR
Letters Patent there is contained a proviso obligin
Henson, by an instrument in writing under my
to describe and ascertain the nature of my sai
manner the same is to be performed, and to caus
in Her Majesty's High Court of Chancery w
next and immediately after the date of the said re
and by the same, reference being thereunto had,
appear.

The first major step forward from the cut-throat was taken in 1895 by a salesman in Detroit, with the splendid name King Camp Gillette. He had written an extraordinary book, *The Human Drift*, in which he proposed that the entire population of the United States, then sixty million, should move to a vast city called Metropolis, built of 40,000 huge tower blocks clustered around Niagara Falls, which would provide hydro-electric power for all. They would eat in vast communal halls, and all work for the same company. Gillette's boss William Painter, inventor of the disposable bottle cap, was scornful of these ideas, and said that if Gillette wanted to become rich and famous he should invent something that people would buy and then throw away. Gillette came up with the idea of the disposable razor blade – a thin plate of steel with two sharp edges that was clamped on to a handle to make a safety razor. This was a superb innovation. Each blade was cheap and could be thrown away when it became blunt and replaced in seconds by a new one. What is more, the blades were so flexible that they could be sharpened by being rubbed to and fro around the inside of a glass tumbler, which significantly improved their useful life.

The patent of King Camp Gillette, with the critical new idea of disposable blades.

Gillette sold 168 disposable blades in his first year of production, in 1903, and 124,000 the next year. When he persuaded the army to issue them to all its soldiers the sales leaped into the millions, and indeed the blades were so successful that they dominated the market for sixty years. They also had the distinction of being one of the first mass-produced products that were designed to be thrown away. Other early disposable objects included bottles, bottle caps and steel pen nibs, but the Gillette razor blades had a much shorter life. We have so many disposable things these days that it is hard to realise what a surprising idea this must have been at the time.

Today, if you ask for a shave in a barber's shop, you may well be shaved with a razor that looks like a cut-throat but in fact has disposable blades clipped into a steel holder. (Surgeons' scalpels use similar technology.) These modern disposable blades are sharper than any edge that can be achieved by stropping a cut-throat and are much less dangerous, since their holders provide a blunt edge parallel to the sharp edge and only a couple of millimetres away; this means you cannot easily cut yourself badly.

The makers of these razors competed for decades to make sharper and longer-lasting blades, and then, in 1975, were caught on the hop by a new upstart manufacturer, who had the temerity to march straight into the marketplace with the ridiculous idea of not just a disposable blade but a disposable razor. Within a few years, disposable razors had taken the lion's share of the market, and now in my supermarket, they are almost alone.

What is more, there is now a great variety of disposables – fifteen different models on the day I checked. Some have single fixed blades, others have two, three or more, and some of these have lubricating strips of polyethylene oxide between the blades. I thought these multiple blades were merely hype and marketing, but there turn out to be good reasons for using them. The idea is that as the first blade cuts a hair, it pulls the hair a little way out of its follicle at the same time; before the hair has time to spring all the way back, the second blade shaves it closer still. The result is that in effect a multi-blade razor will cut the hairs below the normal level of the skin.

Hi-tech disposable razors now sport two, three, four and even five blades.

When you shave with a two-bladed razor, the skin bulges up slightly between the blades. This means that the second blade meets the skin at a slightly different angle than the first, which can be uncomfortable. For this reason, the more blades there are, and the closer together they are, the less bulge there is and the less pressure on any of the blades, so you get a more comfortable shave. The problem is that washing out the gunge and cut hairs becomes more difficult as the gap between the blades decreases. Also, the manufacturing process becomes steadily trickier because if one blade is 40 microns out of alignment, it produces an uncomfortable shave. In practice, each blade is individually spot-welded in place by lasers, while the geometry is held with high precision.

Most good disposable razors have swivelling heads to follow the contours of the skin; the multiple blades don't work unless they are all in contact, and a swivelling head is necessary to achieve this. There are lightweight models in tasteful pink for women, and even razors with built-in vibrators 'to instantly reveal more radiant skin'. Still the selection pressure remains. Razors will appear and disappear, and gradually will get better and better.

The blades themselves are made of complex composite materials, with layers of niobium, diamond-like carbon, chromium and PTFE on top of a base of steel. The edge of the blade is wedge-shaped and is measured in microns, where a micron is 1,000th of a millimetre. The edge is so sharp that a micron back from the edge, the blade is only half a micron thick. I remember hearing someone say in about 1960 that if he won a lot of money, he would spend some of it on the ultimate luxury – a new razor blade every day. Now this is within the range of people who have not won any money, but the twenty-first-century blades are so much better that they scarcely seem to be less sharp after a week's shaving; the average lifespan seems to be about two weeks. In fact, razor blades are the most high-tech of all the mass-produced things in the world. I look forward with interest to future developments in the evolution of this cutting-edge technology.

3 The Outer and Inner You

CLOTHES

Why do I get dressed? Partly to keep warm, since the temperature in Britain is generally too low for naked comfort, partly to hide my body from public display, and partly to make a fashion statement. I usually wear clothes in layers. As I write this, on a frosty day in November, I have a thermal vest made of knitted cotton and a shirt woven from a mixture of cotton and polyester, which looks tidy but is easy to look after, since it does not need ironing.

Pure cotton shirts need ironing to prevent them from looking crumpled. Some people tell me they actually enjoy ironing, and if that is true, I would not seek to stop them, but I don't think that anyone should have to iron shirts. After washing, cotton looks crumpled for the same

Electron micrographs of a cotton shirt collar, before (left) and after washing. Note how washing has removed lumps of dirt and the layer of grease.

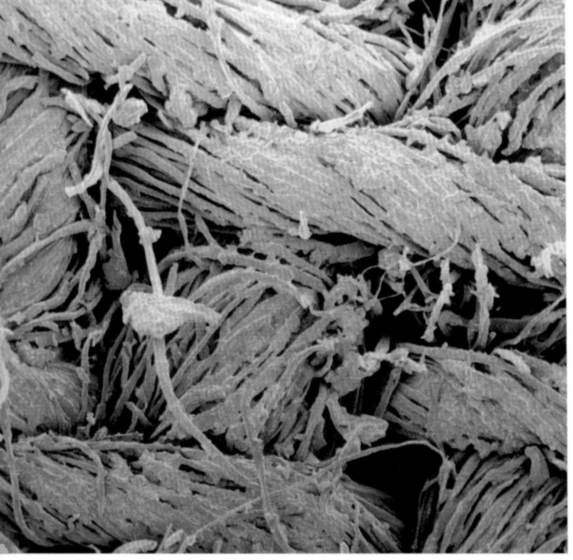

reason as hair needs careful drying: cotton is made of the chemical cellulose, and hydrogen bonds between separate molecules of cellulose will form when the water evaporates. Unless the material is perfectly flat and smooth, the new bonds will form random creases in the material, so that it looks crumpled.

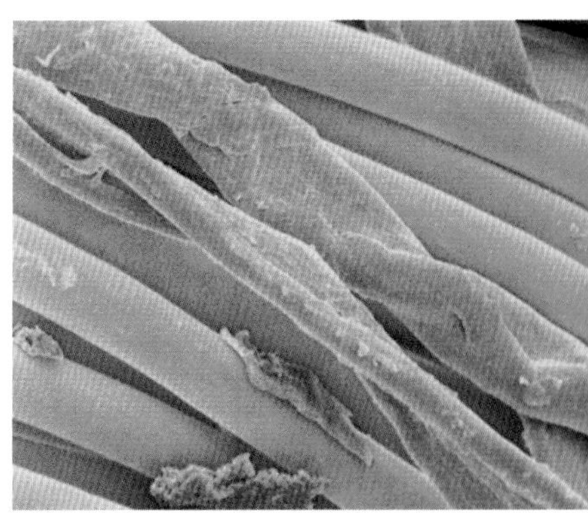

A mixture of polyester and cotton fibres; these will drip dry without crumpling.

When the cotton is mixed with some polyester – typically 10 to 50 per cent – to make a 'poly-cotton' shirt, the shirt is less likely to get crumpled, because the molecules of polyester form fewer and weaker hydrogen bonds, which means fewer rogue creases. However, the downside is that polyester is less absorbent than cotton, which makes it slightly less comfortable. Also, a well-ironed cotton shirt is arguably smarter than a poly-cotton shirt, which cannot hold such firm creases. Shirts made of pure polyester or nylon drip-dry beautifully, but are much less pleasant to wear, mainly because these synthetic fibres do not absorb water and therefore do not wick the sweat away from your skin.

Cotton creases are less severe if the shirt is dried in the wind on a washing line, for it should hang in roughly its 'natural' attitude, and continual gentle movement will discourage the formation of 'rogue' hydrogen bonds. This is why drip-drying is fairly effective, even for cotton. People who travel a great deal on business stay frequently in hotel rooms and also have to look smart; so they often use a neat trick. They hang their clothes overnight on hangers in a warm, damp place like the shower cubicle. The damp air frees any crumples from the clothes, and because they are on hangers, they will regain their natural shape. The reason this works is that water vapour is absorbed by the cotton – it can hold 13 per cent of water in a steamy atmosphere – and this water 'plasticises' the cotton by releasing all the old hydrogen bonds. The fibres can then move until they lie flat again. That is why the steam iron is so effective. The steam plasticises the cotton, just as it plasticises wood to bend (see page 33), and the heat and pressure from the iron encourage new hydrogen bonds to form in

Warm, dry, windy weather is best for drying clothes on the washing line, because the water evaporates quickly and the water vapour is carried away by the wind.

the correct places for perfect sharp creases and no crumpling.

When I plan to do something energetic, such as riding my bike, I like to wear layers that wick away the sweat. Wearing an ordinary cotton T-shirt next to the skin is acceptable on a warm summer's day, but with another layer on top, it can become wet and cold if I take a rest – or even when I am coasting downhill. The solution is to wear a loosely woven base layer that will allow the sweat to pass through but keep a layer of warm, dry air next to the skin. Although it is less satisfactory for a shirt, synthetic fabric such as polyester is good for this base layer, since the fibres do not absorb water and therefore do not get wet and uncomfortable.

As long as I do not have to look smart, I like wearing fleeces on the outside. Modern fleeces are light, warm and breathable; they are easy to wash, and they dry quickly. Some are made from recycled plastic bottles, which makes them environmentally friendly. The American Plastics Council claim that some 35 per cent of plastic soft-drink bottles are recycled, and those made from the polyester PET (polyethylene terephthalate) are chopped into flakes, cleaned and melted. The liquid PET is squeezed through small holes to make long strands of fibre, which are spun into yarn and woven to make the fleece material. Each fleece garment needs about twenty-five 2-litre bottles, and other products include T-shirts, shoes and carpets.

I have a collection of woollen sweaters, which I wear occasionally. In Britain, before about 1750, clothes were largely made of wool, but because there are now so many synthetic materials, wool is largely being replaced. This is partly because woollen clothes are tricky to wash, and without care they change shape when they dry; in particular, the bodies of sweaters shrink, while the sleeves stretch, so that they become progressively less suitable for me and more for a gorilla. This is partly a result of the breaking and remaking of hydrogen bonds between the molecules of the wool, and partly because the fibres slide over one another but won't slide back because of a ratchet effect.

A mixture of wool fibres (looking a bit like the hair on page 33) and smooth synthetic fibres.

The best suits are made of pure wool and are kept in good shape by dry-cleaning. This means removing the dirt and the grease with organic solvents rather than water. Most dirt is mixed with grease or oil, which is why you need something like soap or detergent to wash it off with water (see page 28), but grease and oil are soluble in organic solvents, so you can simply rinse clothes with these organic solvents and they will remove the dirt directly. The solvent most commonly used for dry-cleaning is tetrachloroethylene, also known as perchloroethylene, or perc. Organic solvents like perc do not form hydrogen bonds, so they do not change the shape of the garment as they remove the dirt.

I don't particularly enjoy dressing up, but I do like unusual and brightly coloured clothes. There are photographs from long ago showing my preference for bright colours, but I recall one incident from about 1979, when I was a researcher at Yorkshire Television.

In those days, I normally wore grey trousers, with a shirt, tie and jacket, and I parted my hair on the left. One Friday afternoon, no one in the office was doing much work, and we began to discuss whether people could completely change their personalities. Someone said that you were defined by your clothes, and after some banter we all agreed that we would wear totally different clothes on Monday morning.

After rummaging through my wardrobe, I put on some white cricket trousers, a coloured open-necked shirt and a sweater, and combed my hair straight back, without a parting. I felt a complete idiot, and sneaked into the office quietly and quickly to see what everyone else had done.

No one else had bothered.

I was on my own.

Although I felt exceedingly foolish, the others were impressed at my metamorphosis and said that I looked ten years younger. Ever since then I have tried not to take my appearance – or myself – too seriously. Some people want to fit in with the crowd and therefore wear the standard uniform. I would rather be an individual.

Occasionally, I have to wear a suit, and I now have three asymmetric suits, with alternate panels made from materials of different colours. One is two-tone red, one two-tone blue, and one is a striking mixture of rust-red and elephant-grey, for use on special occasions.

Sometimes I have to wear a suit, but I try to avoid being too conventional.

SHOES AND SOCKS

Bare feet are uncomfortable on rough ground, and I usually wear socks and shoes. Indoors at home, I like to wear slippers or mules, but elsewhere I prefer my multicoloured desert boots, or 'Noddy boots', as one friend described them. These were made for me by the Bristol Handmade Shoe Company, who offer only a limited range of styles, which appeals to my minimalist instincts. Each shoe is made from just four pieces of soft leather stitched on to a composition sole. The chap who makes them first got me to stand on a piece of paper and drew round my foot with a pencil, to get the right size and shape; now he keeps the outline on file in case I go back for a new pair, which I probably will; they have made me three pairs so far.

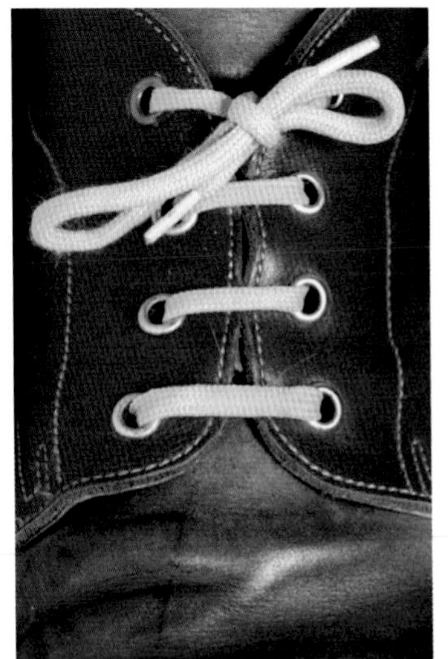

Shoelaces – and most clothes – are utterly dependent on friction.

Roman soldiers wore hobnailed sandals. The hobnails ensured that the sandals gave a good grip in the ground and lasted a long time. The open construction of the sandals must have kept those Roman feet cool in hot weather and allowed any water to run off if they had to march through puddles or wade through streams, but they must have been perishing cold in snow.

Many modern shoes are made from nylon and other synthetic materials, which can be strong, waterproof and made in any three-dimensional shape. Traditional shoes, however, are made from leather, which is tough, rigid enough to keep its shape and yet soft enough to be comfortable on that rather intricate object the human foot. Leather is also fairly breathable – that is, it allows sweat to evaporate, so that the feet don't get too soggy.

When asked what colour I would like, I said blue and green and red, in this order. Some people have both shoes the same colour, but this seems to me to be a sad waste of opportunity, when we could be using our shoes to brighten the world. This particular pair has so far coped with snow in the French Alps, dust in the desert of Kenya's Rift Valley and several television series in Britain.

These 'Noddy boots' are fastened with old-fashioned shoelaces, which are just bits of string with non-fraying ends. Shoelaces have lasted remarkably well, and hold their own even against elastic-sided slip-ons and such modern fasteners as zips and Velcro. Laces, usually tied with a half hitch and a bow on top – technically, a double quick-release granny – remain tied because of friction. All knots, and indeed all clothes, are held together by friction, which we regard as tiresome when we are trying to slide heavy furniture about, but without it we should all be naked. Perhaps that is an exaggeration, but the vast majority of clothes certainly need friction.

Recently, while visiting a naval establishment, I was reprimanded for having improper footwear, so now I make a point of sporting port and starboard socks, so that they can't complain. Boats at night carry riding lights – red on the port or left side, and green on the starboard or right. These tell others which way the boat is travelling. Wearing a red sock on my left foot and a green one on my right saves me some time in the morning, since otherwise I should have to worry about which sock to put on which foot.

Also, should I ever get athlete's foot on one foot, it would not be spread to the other by my socks, whereas this would be a danger if the two socks were indistinguishable and I were to wear them for more than one day.

Apart from the brightness, I like my socks to be soft, elastic and long enough to tuck my trousers into when I want to ride a bike. They are generally made of a mixture of cotton (strong and absorbent, but not elastic), wool (soft and warm, but not hard-wearing) and nylon (strong, hard-wearing and elastic, but not absorbent and therefore liable to get sweaty).

Having both shoes – and socks – the same colour seems a waste of opportunity.

I have other forms of footwear too – trainers for the gym, hiking boots for long walks on rough ground, wellington boots (allegedly popularised by the Duke of Wellington, the man who beat Napoleon at Waterloo in 1815) for working as gardening assistant, and grotty old sandals for the beach. I also have dark- and light-blue socks, to go with my two-tone blue suit, and support stockings, provided for me during a hospital stay, which I wear on long flights, since I am told that by constricting the blood vessels near the skin, they diminish the chances of deep-vein thrombosis, which is a slight hazard for frequent flyers.

While filming a programme about what the Meso-Americans did for us a few years ago, I had one foot painted with black latex, a primitive form of rubber apparently used by the Incas. I then walked through puddles to see what happened. The unpainted foot got wet, and the painted one probably stayed dry – certainly the water ran off it rapidly – but I could not feel the difference, since basically both feet felt cold. I was not convinced that these 'contact wellies' were useful – and weeks went by before I managed to remove all the black from around my toenails.

KEEPING TIME

Sadly, I am not a punctual person; I tend to be late for things. When I was twelve, during some weeks I was responsible for ringing the school bells to summon all the boys for lessons, or lunch or whatever was next. Even though I was the proud owner of a new watch, I kept missing the moment and ringing the bell five minutes late.

At home, I have a grandfather clock that my dad bought for £1 in a sale in Yorkshire. Upstairs in my room is a fine old American clock that was given to my dad by his great friend George Lyttelton. George had been one of my dad's teachers, and when they met at a dinner party in 1955, George complained that no one wrote to him any more. Dad put pen to paper that weekend, and from then on they wrote to one another every week until George died in 1962. Those 600 missives have been published as *The Lyttelton-Hart-Davis Letters*, and my clock is a constant reminder of the two old literary codgers.

I don't like wearing a watch on my wrist, for my skin becomes sore – I am probably allergic to nickel, which is often a component of watch-strap buckles. So I wear a PopSwatch. Each one comes with a wide rubber strap, but I throw this away and clip the Swatch to my shirt using the plastic clip provided (see picture on page 141).

When I first heard about this, I liked the idea, and the next Swatch shop I passed happened to be on Fifth Avenue in New York City, so I went in and bought one. Since then I have acquired about six, in various designs, and I still have about four that work. These splendid devices are cheap, but keep excellent time; they run for a year or more on a battery. Occasionally, they fall off, but usually I notice and pick the pieces up again. PopSwatches are sealed to 30 metres, which corresponds to a

This American Waterbury clock belonged to my dad; it has a pendulum and weights, and needs winding every day.

water pressure of about three times atmospheric pressure. I am unlikely ever to dive to 30 metres, but this feature is useful nonetheless, for every now and then, when I do the weekly wash in the washing machine, I leave the watch on the shirt by mistake. It comes out very clean and still working, but ten minutes slow, for it seems not to work during the fast spin, when its local gravity must be seriously disrupted.

FURNITURE

Each summer I try to escape for a week or so to Clissett Wood in Herefordshire and make a chair or other useful object from green wood using only hand tools, mainly the draw-knife, the pole lathe, and the adze. I have so far made two chairs, a bench, a table and a folding tray. I am learning this skill under the excellent guidance of Gudrun Leitz, who runs courses all through the summer and encourages experimentation, but brings me and her other pupils back into line when we are in danger of ruining our work (see www.greenwoodwork.co.uk for details).

I made these chairs from green wood, the first on the left and the second on the right. They have been in daily use for several years.

The first chair I made was a Windsor chair; the seat is elm, and all the rest is ash – the legs, the three stretchers that hold the legs apart, the spindles up the back, and the comb across the top. The legs came first. You take a length of ash trunk about 10 centimetres thick and split it lengthwise into four. Each quarter will make one leg. To make each leg from the whole of a thinner branch would not be so good, because the centre of each branch can be weak. Therefore you have to shape each quarter-cylinder into a cylinder, using first a side-axe, second a draw-knife on a shave-horse, and finally a pole lathe.

Working on a chair leg at the pole lathe.

The side-axe is a short-handled chopper, with the blade set not symmetrically but on one side of the head and the handle twisted away from this side. This makes it ideal for slicing off the rough corners of the split branch. The wooden shave-horse has a seat at one end and a cunning lever system at the other for clamping the rough cylinder with one end pointing slightly upwards towards your abdomen. You then hold the draw-knife with both hands, lay the blade on the leg-to-be and pull towards you, thus shaving off a strip of wood. Continue doing this, rotating the wood progressively and turning the far end towards you from time to time, until you have made a fairly smooth cylinder. Then you are ready for the pole lathe, which is a beautiful instrument.

Mount the rough cylinder between the spindle points so that it is horizontal and at chest level, having first taken a couple of turns of cord around it. One end of this cord goes down to a treadle on the ground, the other up to the end of a bendy pole above your head. The pole, which gives the pole lathe its name, is propped up on a support a couple of metres in front of you, and the far end is anchored to the ground. When you are all set, you press down with your right foot and the cord wrapped round the wood makes it spin. Using both hands, you rest a chisel on a horizontal bar about a centimetre away from the rough cylinder and gently ease the blade against the wood, where it removes a ring as the wood spins. Then you pull the chisel back a few millimetres, lift your foot, and the spring of the pole pulls the cord up and spins the working piece back the other way, ready for the next stroke. The springiness of the pole is what returns the working piece to the right place to begin again.

To begin with, the surface is rough and the chisel jumps and jerks, but after a few minutes you have smoothed the working piece considerably, and after perhaps ten minutes even I can make something that begins to look like a chair leg. Then further work with chisels of various shapes can yield elegant variations, with whatever beads and grooves you want.

This is skilled craftsmanship. Learning the basics takes perhaps an hour, but the finer and more intricate I want it, the trickier it becomes, and I often get myself into trouble and need rescuing. Despite this – or perhaps because of it – working with the pole lathe is utterly absorbing. I have to concentrate totally on what I am doing or I make terrible mistakes and gouge lumps out of my developing cylinder. I have to watch what is happening to the wood, noting the grain and the thickness, and watching out for knots. I have to get it to the right thickness, say 31 millimetres tapering to 25 millimetres. And I have to do this for each chair leg in succession, while standing on one leg of my own, which is not something I often do in my normal life. I should also try to get them all the same, although in my first chair the legs all finished slightly different, owing to my incompetence, so in my second I deliberately made them all different to forestall criticism (see page 62).

Clissett Wood has no buildings and no engines. As I stand there surrounded by trees (and other woodworkers), there is only me and the wood and the ssshhh…ssshhh…sound of the chisel on the wood at each working stroke of the lathe. This is soothing and almost spiritual. I do not quite get into flow as I can in my photography (see page 140), because I am surrounded by others, and by the sounds of their lathing and splitting and side-axing – and occasionally their whoops of delight when something goes right or groans of frustration when it doesn't. Nevertheless, I am away from my phone and my email and my computer. I have to switch off that side of my life and think only about the wood, which must be a Good Thing. What's more, taking physical exercise all day is hard work, especially when standing on one leg, and by evening I am pleasantly tired and relaxed.

Woodworking with hand tools in 1900.

For my garden bench, I turned the legs from oak, which is surprisingly different from ash. It isn't significantly harder, contrary to my expectations, but the shavings come off not in long strips but in little chunks, which are a distinct irritant; they have a noticeable smell, and eventually they affect your skin. Gudrun turned a thousand oak spindles for the Globe Theatre in London and suffered a tedious rash as a result.

For each of my chairs, the seat started life as a rough plank of elm. I sawed the corners off to make approximately the right shape as seen from above, hollowed the seat slightly with an adze, which is a vicious curved blade on the end of a wooden handle, and then smoothed the surface mainly with an inshave and a spokeshave. This is hard and tedious work, again demanding great concentration. Elm is tough wood, with weirdly twisted grain, which makes it extremely difficult to finish without lifting unwanted slivers. However, the grain is fascinating, and as you expose it, the seat becomes steadily more beautiful. I must confess I found the adzing exceptionally hard work, and I had a good deal of help with it, but I did the subsequent work myself.

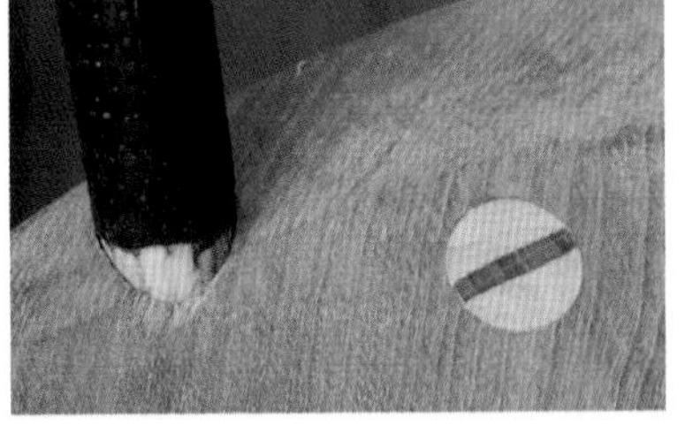

Detail from the second chair, showing the sweet chestnut whittled down and socketed into the seat, and the leg coming through from below and held in place with a dark oak wedge.

While you are working on the seat, the legs and the spindles go off to be dried, ideally for two or three days in a slow oven. They shrink a little as they dry, and because they are taken not from the centre of the branch but from off-centre, they shrink asymmetrically into ovals. These need to be turned down to true cylinders on the pole lathe, so that they fit into the sockets that you then drill into the elm seat. When all is ready, and the holes have been drilled at the angles handed down by generations of chair-makers, or chair-bodgers, you saw slots in the tops of the legs, smear the newly rounded ends with a hint of glue, hammer the stretchers into the legs and the legs into the seat, saw off the protruding leg-ends flush with the seat, and hammer thin oak wedges into the slots so that the legs cannot come out again.

The back of my second chair was made from a piece of sweet chestnut, steamed and bent round a former. The centre Y-piece is of hazel, carefully selected, cut to length, and the ends whittled to 12 millimetres to fit into 12-millimetre holes drilled in the sweet chestnut. The bark did not strip off the sweet chestnut when we bent it, which is unusual, and I decided to leave it on both this and the hazel, to make it look as rustic as possible. Both chairs remain strong and are in daily use, and I am happy with them.

My two-seat bench, trestle table and folding tray.

My oak-and-elm garden bench has survived several years of weather and use, and shows no signs of collapse. It has now been joined by an elm table, shaped largely with a draw-knife. The legs, of sweet chestnut, I made into trestles, which stand up on their own, and the tabletop simply rests on them; it is not fixed in any way. This lives in the garden in the summer, and in the garage in the winter. I oiled and varnished the top to bring out the grain, whereas I did not treat the bench in any way and it has turned silvery grey. Tables are almost as useful as chairs and again tend to come in just a few basic designs. Their function is to provide surfaces at heights convenient for people to reach easily while sitting on chairs – to reach a keyboard, drink or food.

Furniture has a much longer history than written history, and we don't know much about how it started. The Stone Age houses at Skara Brae on Orkney, five thousand years old, have stone-walled beds on the ground, stone dressers and lumps of stone to sit on. There are no tables, but then to make a table from stone is not simple. In many later civilisations, there were surely wooden tables and other furniture, but wood does not usually last for thousands of years, so we lack the evidence. We do know that the Ancient Egyptians had furniture that looks much like ours, since some was preserved in tombs and depicted in paintings. They were fond of folding X-frame stools, but they also had wooden beds, tables, chairs and other simple items. Grand people had furniture made from gold, silver, bronze, marble and other precious materials. Today we rarely use precious metals for furniture, but we do use iron, chrome, glass, leather and all sorts of other materials to make our lives more comfortable and – perhaps – more elegant.

Meanwhile, at the cheap end of the market, vast quantities of furniture are now made from medium-density fibreboard, or MDF, which is made from wood fibres glued together. The advantages of MDF are that it is cheap, flat, smooth, comes in large sheets – much wider than natural planks cut from trees – and has no grain, so it can be sawn through in any direction. It can be drilled and screwed together, although screws do not hold as well as they do in real wood. The other disadvantages are that it is dense, so that large sheets are heavy; it can crumble at the edges, and it does not look or feel like real wood. Also, it swells when wet, so it needs to be covered, either with paint or with a plastic surface such as melamine.

Handmade furniture can be beautiful, but it is always going to be expensive. Mass-produced furniture can be both cheap and surprisingly attractive, not to mention immensely popular. One of the most successful furniture retailers is IKEA, and the Swedish man who founded the chain, Ingvar Kamprad, is said to be one of the richest men in the world.

BREAKFAST

During the night I eat nothing – in other words, I fast – and I need to break that fast to gain energy for the day ahead. At home, I usually have some wholegrain cereal with milk and some fruit – banana or raspberries or chopped-up nectarine – plus some fruit juice and a cup of tea. Actually, the very first thing I have is a cup of tea (see page 25), but I don't count that as breakfast, which is a sit-down meal in the kitchen or the garden.

The idea of starting the day with processed cereal was invented and popularised a hundred years ago by an evangelical American doctor called John Harvey Kellogg, who was convinced that most human diseases arise from constipation. He performed 20,000 bowel operations and claimed that he never found a healthy bowel. He asserted that constipation allows all sorts of toxins to build up and poison the body – and incidentally that constipation is the primary cause of masturbation – and if only people could avoid constipation, they would all be as fit as fiddles.

Breakfast cereals were invented by John Harvey Kellogg as a cure for constipation.

He went on to say that the way to stimulate regular bowel movements was to give up the traditional American breakfast of ham and eggs, and instead eat plenty of fibrous cereal. At the Battle Creek Sanitarium in Michigan, with the help of his brother, Will Keith Kellogg, he developed first granola, then cornflakes, and followed on with a sequence of other health foods, including soya products ('non-meat meat'), coffee substitute and peanut butter.

Among Dr Kellogg's patients were John D. Rockefeller, Thomas Edison, Henry Ford and George Bernard Shaw, so he must have been a persuasive man. He certainly turned Battle Creek into the breakfast-cereal capital of the world, and his name lives on, immortalised on countless cardboard packets.

Often during the winter, I eat porridge in place of cereal. Porridge is generally made of rolled oats and needs to be slightly cooked to make it palatable, preferably with water or milk. It can then be eaten on its own, with salt, with milk, sugar, maple syrup, or with fruit. I like it with a little milk and some sultanas, seeds and nuts put in during the cooking so that they soften, and perhaps a banana or some other fruit on top.

According to the *Oxford English Dictionary*, porridge, which can also be spelled 'porage', 'porradge' and in a dozen other ways, comes from the word 'pottage', meaning 'soup', and in Scottish and English dialects is usually treated as a collective plural and called 'they' – rather like a herd (of cows) or a board (of directors) –'having a few porage made of the broth of the same beef, with salt and oatmeal' (Lever, 1550). Scottish folklore has it that porridge should always be stirred clockwise, using a straight stick (or 'spurtle') held in the right hand, and eaten standing up with a bone spoon; but I confess these traditions are not strictly followed in the Hart-Davis household. Some people take porridge rather more seriously than I; there are even world porridge-making championships every year in Scotland (see www.goldenspurtle.com for details).

Porridge can be made with almost any cereal grains. They just need to be cracked open, soaked, and then cooked for a short time in water in a pan on the fire – no need for an oven or special equipment. Porridge must have made simple food for early farmers ever since the Bronze Age, when cooks first had pots, and it is still a staple food in many parts of the world: Kenyan women, for example, make a porridge called *ugali* from maize flour. Cooking porridge is not entirely trivial, however; in 1573 William Tyndale, the first person to translate the Bible into English, wrote in *The Obedience of a Christian Man*, 'If the porage be burned... we say the bishop hath put his foot in the pot.'

For Scottish agricultural workers (or 'farm servants'), porridge used to be a staple diet. A cook would go two or three times a week into the bothy where they lived and make a huge cauldron of the stuff, which was then stored in the drawers of the kitchen dresser, where it solidified. Each man would cut a thick slice every day and take it for his lunch (or 'piece') in the fields.

Porridge played an important role in the diet of Scottish agricultural workers in the sixteenth century.

We digest our food by hydrolysis; that is, we eat long molecules of carbohydrate, such as starch, and hydrolyse it – react it with water, using enzymes, to break it into smaller molecules, such as glucose, which we can absorb into the bloodstream. Likewise, we hydrolyse protein into amino acids, which we use to build our own muscles. The muscles operate by reacting or 'burning' glucose with oxygen, which is brought along in the bloodstream. Each muscle has its own store of glycogen, which is a sort of animal starch and acts as a store of glucose. The glycogen store will last for about six hours before it needs to be topped up with more glucose, which means that glucose tablets do not supply instant energy; they may taste good, but they will take several hours to have any effect on the muscles.

One advantage of porridge is that the carbohydrate is locked more tightly into the oats than it is into cornflakes and other highly processed cereals, so that glucose is released more slowly and steadily into the bloodstream. Sugary cereals provide an instant punch, but release all the carbohydrate too soon; so you may get hungry again by mid-morning. A recent study by Professor Jeya Henry with a group of Oxford schoolchildren aged eleven to thirteen found that their appetite for lunch depended on what they had eaten for breakfast. After eating cornflakes, which have a high glycemic index – it rapidly raises the level of glucose in the bloodstream – they became hungry by midday and ate a large lunch. Those who had eaten porridge or All Bran, which release glucose much more slowly, felt less hungry and ate less lunch. Perhaps this finding will help me lose weight; in future I shall continue to eat porridge or All Bran – or cottage cheese and fruit, which I am told is another good option. Even adding a small amount of sugar does not do much harm.

What I shall definitely do is find out more about this glycemic index, which is explained on a helpful website – www.glycemicindex.com – and is a score for carbohydrates. Those that produce high blood-sugar levels after you eat them get high GI scores – up to a hundred – while those

Cereal	GI
All Bran	39*
Bran Flakes	74
Coco Pops	77
Cornflakes	82*
Frosties	55
Rice Krispies	82
Shredded Wheat	75*
Special K	69*
Weetabix	74*

(* = average figure)

that cause a gradual change in blood-sugar level have low GI scores. Sticking to a low-GI diet seems to lower your risk of suffering from obesity, diabetes and coronary heart disease.

I looked up the GI values of some common breakfast cereals and was rather surprised by various things. First, All Bran is the only one that has a GI lower than fifty. Second, sugar does not seem to matter too much, since Frosties score much better than cornflakes, even though they are effectively cornflakes coated with sugar; similarly, Coco Pops are apparently better than Rice Krispies, which taste as though they are all air.

I ride a bicycle almost every day, and I look at some of the cycling magazines, which often include articles about breakfast. Cyclists who want to ride all morning need slow release of energy. Some believe in having masses of cereal and toast for that early boost and carrying bananas for a top-up later on. Others believe in a full cooked breakfast of bacon, eggs, sausages and the rest, because although they feel full and heavy to begin with, that feeling wears off, and they do not get hungry mid-morning. Porridge may be the answer, but in fact because digestion is slow, what the cyclists eat for breakfast is probably more important for psychological than physiological reasons.

A super-healthy breakfast of All-bran, cottage cheese and fresh fruit, plus cloudy apple juice – allegedly better for you than the clear variety.

There is wide cultural variation in breakfast. I am not particularly enthusiastic about the northern European version – bread with cheese, ham or salami-style processed meat. Indeed, I prefer to wrestle with a Japanese breakfast, which may include rice, pickled vegetables, raw fish,

fermented bean curd (*natto* – not to everyone's taste, even in Japan), and perhaps a raw egg. The egg is lightly beaten with soy sauce and stirred into a bowl of hot rice, where it cooks slightly to form a glutinous mass, but this can present a bit of a challenge, especially using chopsticks, or after a late night. One of the greatest breakfast contrasts, however, is displayed across the English Channel.

A full English (or Scottish or Irish) breakfast is an absurdly complicated meal, usually offering fruit juice followed by bacon, eggs (cooked any way), sausage, black pudding, mushrooms, baked beans and

Contrast between a northern-European style breakfast and the absurdly complicated 'full English'.

fried bread, or kippers, or haddock with a poached egg. Even in the cheapest hotel, the table is cluttered with plates, bowls, knives, forks and spoons, with toast, bread or buns, butter and marmalade, honey, and jam, salt, pepper, mustard and perhaps red and brown sauces, cups and saucers, milk jug and sugar – a nonsense of complexity. In a French motel, I watched a man dipping his croissant into his bowl of coffee while holding a newspaper in his other hand. One piece of crockery, no cutlery, and nothing else on the table: what elegant simplicity.

Americans take the full English breakfast and add even more options: steak, hash browns (fried potato), smoked salmon, bagels, cream cheese, fresh fruit and waffles or pancakes with syrup. But if you really want variety, try an international hotel in Japan, where you will find every single one of the above ingredients on tempting display.

I never cook a big fry-up, but on Saturdays I make boiled eggs for Sue and myself. I take the eggs from the fridge, put two for myself into a small pan containing warm water from the tap, and put it on the burner. When the water boils, I put Sue's egg in; I leave the water to boil for exactly five minutes and then remove all the eggs. This leaves hers soft in the middle, while mine are just hard. I have a cunning device patented by engineer Duane Chadwick from Utah called 'The Good Egg', which looks like a fat thermometer and has the same sort of heat-transfer properties as a real egg, so that it tells you when your egg is perfectly cooked – and it allows for the size of the egg and for whether you want it hard, medium or soft. I like the idea of this device, but the fact is that I have found the correct times by trial and error, and a timer is easier to use.

Patents for Duane Chadwick's 'Good Egg', which allows you to boil your egg perfectly every time, and Thomas Gaddes's automatic egg boiler.

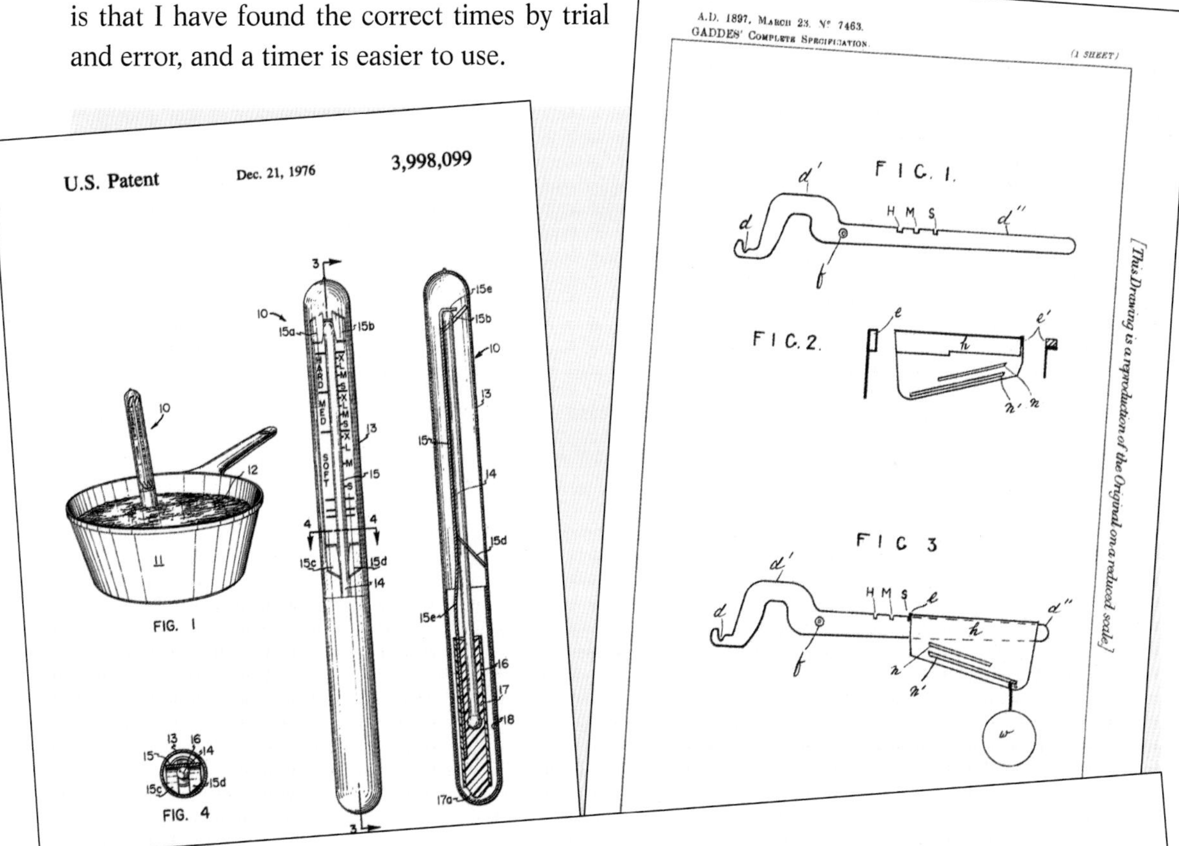

A.D. 1897, MARCH 23. Nº 7463.
GADDES' COMPLETE SPECIFICATION.
(1 SHEET)

[This Drawing is a reproduction of the Original on a reduced scale.]

U.S. Patent Dec. 21, 1976 3,998,099

Nº 7463  A.D. 1897

Date of Application, 23rd Mar., 1897—Accepted, 22nd May, 1897

COMPLETE SPECIFICATION.

An Improvement in an Automatic Egg Boiler.

THOMAS GADDES, M.D., Dental Surgeon, 104, Station Parade, Harrogate, Yorks, do hereby declare the nature of this invention and in what manner the [illegible] to be particularly described and ascertained in and

The Gaddes egg boiler as constructed for TV series *Local Heroes*.

In addition, I have a modern reconstruction of the automatic egg boiler patented by Dr Thomas Gaddes, a dentist from Harrogate in the north of England. Gaddes must have been either absent-minded or worried that he might suddenly be called away to deal with a patient, for he designed this cunning device to remove his egg from the boiling water after precisely the right time, and so prevent it from being too hard. The original 1895 device was rather complicated, and when he realised he could make the machine simpler, he took out a second patent in 1897; this is the version I have. Basically, the egg is suspended in a cradle from a beam with a counterweight on the other end. Above the egg is a small reservoir of water with a hole in the bottom, reminiscent of the *klepshydra* (see page 11). As the water leaks out, the water reservoir becomes lighter and the beam begins to tilt. At the critical moment, a subsidiary weight slides along a wire, tipping the balance suddenly in favour of the counterweight, and the egg is snatched from the boiling water, cooked to perfection.

On top of a high mountain, an egg would have to be boiled for a long time to become hard-boiled, because the lower air pressure at high altitude would cause water to boil at a lower temperature. I regret to say that I have never tried boiling an egg at high altitude so I don't know how long it would take. I should be glad to hear from any egg-boiling mountaineers with practical experience.

Even in Britain some people prefer a simple breakfast – nothing more than coffee and a piece of toast. Toast carries a modicum of carbohydrates and all its own folklore. When I was a lad, we used to make toast by skewering a piece of bread on a toasting fork and holding it in front of an open fire, preferably when the fire was going well, with lots of glowing embers and not too many flames. We had to take care not to scorch our hands or faces, but when we got it right, the toast was wonderful.

Few Westerners have open fires today, but should toast be grilled or made in a toaster? Buttered or dry? Slathered with marmalade or with marmite and tahini? For guidance on toastology, you might like to look at the website www.drtoast.com, which contains a wealth of eulogies, recipes, articles and even haiku in praise of toast, including this one, entitled 'Constellation':

scrape into the sink
constellation of black stars
a toast neglected

The first American patent for a toaster was taken out in 1909, and the first pop-up toaster appeared in 1926. Today, 75 per cent of American homes have toasters. There are plans to build a toaster museum, outlined in www.toaster.org, and you can visit the website www.toastermuseum.com, which has a host of other toastabilia.

At my first boarding school, at the age of nine or ten, we used to get up to various sorts of mischief in our dormitory after lights out, and one night we decided to have a 'midnight feast', which actually took place around 8.30 or 9 p.m. One of the boys

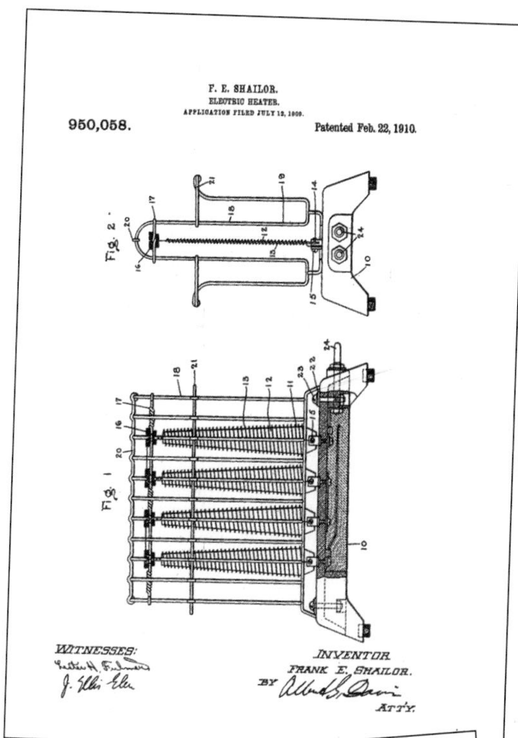

The first patent for an automatic toaster.

UNITED STATES PATENT OFFICE.

FRANK E. SHAILOR, OF DETROIT, MICHIGAN, ASSIGNOR TO GENERAL ELECTRIC COMPANY, A CORPORATION OF NEW YORK.

ELECTRIC HEATER.

Specification of Letters Patent. Patented Feb. 22, 1910.

950,058. Application filed July 12, 1909. Serial No. 507,030.

To all whom it may concern:

which is mounted the high resistance conductor 13. The mica support is fixed in a ... the base. The claimed we could make toast by heating bread over a candle, and we did try, breaking several school rules in the process. The result was clammy bread, slightly warm, with a layer of waxy soot on the surface – utterly inedible. Please don't try this at home.

However you make your toast, it needs to be done quite slowly, over a few minutes, so that the outside surfaces become crisp while the inside gets comfortably warm but remains damp and soft – unless you want the totally crisp melba toast, usually made by making normal toast, slicing it down the middle and then toasting the inside surfaces again. For a television programme, I once tried using a blowtorch on bread. This took only about thirty seconds, but the result was cold, damp bread with a blackened surface; the taste was horrible.

Be careful not to do what I once did as a student. In my room, I had a two-bar electric fire with a large reflector to throw the heat into the room. I bent a wire coat hanger so that it would hang over the top of the reflector with one end in front of the red-hot elements. Then I impaled a piece of bread and hung it there to toast automatically. Unfortunately, someone knocked the coat hanger, which swung in so that the wire end touched the element, thus providing a short circuit to earth. There was a bit of a flash, and although no serious damage was done, I prefer to forget the rest of the event.

VISITING THE LAVATORY

In my youth, I was always told that I should go to the lavatory after breakfast – perhaps this was a dim echo of the influence of John Harvey Kellogg – and at my prep school, as soon as breakfast was done, we were actually herded into the lavatories – four cubicles each side of a narrow corridor – while one of the teachers ticked our names off a list.

What accumulates in the colon, or large intestine, is a mixture of indigestible food, especially fibre from vegetables and fruit – and that breakfast cereal – the remnants from dead blood cells and vast numbers of bacteria that have been feeding on these remains. When fresh and soft, faeces also contain some 70 per cent water. The colour of faeces comes from the breakdown of the old blood cells, which form a brown chemical called bilirubin, which is closely similar to biliverdin, responsible for the green colour of some eggshells.

Kellogg and many subsequent doctors have warned that if these bacteria and their products are allowed to remain in the colon for too long, they may be reabsorbed into the bloodstream and do the body harm. That is one reason to encourage regular defecation. Another reason is that if faeces accumulate, they tend to harden, since the water in them is gradually absorbed by the walls of the bowel. Hard faeces make defecation more difficult and you may have to push long and hard. This may over time lead to local problems such as haemorrhoids (piles) and also increased blood pressure with concomitant risk of stroke. Regular defecation is therefore a Good Thing, but 'regular' varies greatly from person to person. Some people – about a third of the population – crap comfortably once a day after breakfast, some go two or three times a day; for others, 'normal' means every other day, or even once a week.

I understand that the lavatory systems in spacecraft were so off-putting that at least one astronaut refused to eat any solid food in order to avoid defecating, but this will not work. You have to dispose of the bilirubin, the bacteria and the water, and even on a liquid diet you have to crap. However, astronauts are normally given a zero-fibre diet, since the weight of faeces does depend on the amount of fibre in the food.

I phoned NASA to ask about lavatories in space, and after a long pause the woman at the other end said, 'We do not have lavatories; we have waste management.' Luckily, I also had the chance to talk at length to an astronaut, and I learned about how they tackled the problems of excretion in zero gravity and a confined space.

In the early days, the astronauts wore absorbent underwear like diapers, or nappies. These 'intimate-contact devices' were claimed to be functional, although difficult to use, time-consuming and messy. By the end of the 1960s missions were lasting several days, and a 'waste-collection system' (WCS) was designed for use by both men and women. One thing they did have was a vacuum – since space itself is a vacuum. So they urinated into the soft triangular nozzle of a sort of vacuum cleaner, which removed the liquid and squirted it out into space. This sounded fine, but in practice the nozzle, designed to fit both sexes, did not fit anyone well. Also, the vacuum was not quite strong enough, so that whoever went to use the nozzle found it still wet from the last person.

There was also the slight problem that the urine tended to hang around the ship and travel along with it as a cloud of ice particles. Some of these difficulties are alluded to in the dramatic movie *Apollo 13*.

To sit on the WCS, the astronaut extended the privacy curtains, lowered their trousers, and then wriggled on to the seat, with thighs under the restraining bars and feet in the foot restraints, since this is one place you do not want to float away from in zero gravity. They would then push forward the operating handle on the right to open the sliding lid just below the seat and to switch on the airflow fan. This pulled air down through the WCS but also focused eleven jets of air inwards at the point just below the anus where the solid emerged, which had the effect of actually pulling it away. On earlier missions, astronauts had found that without help from gravity the faeces stuck to their bottoms. The air-jets helped, but unfortunately they tended to be icy cold.

In the *Skylab*'s WCS, the faeces were pulled into a 'slinger', which worked like a top-loading spin-dryer. Vanes spinning at 1,500 revolutions per minute shredded the solid and spun it out to the wall of the drum, where it was deposited as a thin layer. This gradually dried out, since closing the lid also opened the vent valve, exposing the drum to the vacuum of space. This worked, slightly too well. After a few days, small pieces of dry solid began to flake off, and some escaped into the cabin. This would not have mattered, except that the astronauts could not resist flicking their weightless peanuts about, and the time came when they could not tell what was what, except by taste...

Medieval doctors were sometimes called 'piss prophets', because they tried to diagnose disease by inspecting the patient's urine.

There were other problems. The excrement tended to build up unevenly in the drum, so every day one member of the crew had to put on a rubber glove and spread it around so that it would not block the system. And each astronaut was allowed to use only one sheet of paper, for fear of preventing the drying process. These problems were later solved by containing the faeces in bags that collected all the solid and liquid, and allowed the air to pass through. A new bag was in place when you sat on the WCS. When you had finished, you lifted the seat, put a lid on the bag and switched on the compactor to compress the bag.

For those of us not in space, what matters most for sheer comfort is the consistency of the product, measured according to the Bristol Stool Form Scale, which ranges from one (small, hard lumps like nuts) to seven (entirely liquid). The most comfortable is four, which is like a smooth,

THE BRISTOL STOOL FORM SCALE

Type 1	Separate hard lumps, like nuts
Type 2	Sausage-like but lumpy
Type 3	Like a sausage but with cracks in the surface
Type 4	Like a sausage or snake, smooth and soft
Type 5	Soft blobs with clear-cut edges
Type 6	Fluffy pieces with ragged edges, a mushy stool
Type 7	Watery, no solid pieces

soft sausage. The Romans had many household gods and goddesses – even one for the lavatory, who was named, rather unromantically, Cloacina, after the Latin word for 'sewer', *cloaca.* Here is a prayer to her, essentially a plea to score a four on the Bristol scale:

O Cloacina, goddess of this place
Look on thy servant with a smiling face.
Soft and cohesive let my offering flow –
Not rudely swift, nor obstinately slow.

Miniature portrait of Victorian plumber Thomas Crapper, painted by his niece Edith.

Meanwhile, both the software and the lavatory itself, the hardware, are subjects we are scarcely allowed to talk about – in polite company, they are taboo. The words we use are all euphemisms: 'lavatory' means a washing place, a basin or a bath; 'toilet' means washing or personal preparation; 'bathroom' is a room with a bath in it; 'cloakroom' is a room for storing coats; 'bog' is a swampy piece of ground; 'loo' may be short for 'waterloo', from 'water closet', which means a cupboard with water inside, but 'loo' could also be short for 'regardez l'eau', or 'gardez l'eau', which is what the Tudors shouted to warn the people in the street, as they emptied their chamber pots out of upstairs windows. The same forms of polite avoidance persist in other languages: '*les toilettes*' in French, '*Toilette*' in German and '*o-tearai*' ('hand-washing place') in

Thomas Crapper's grave in Elmers End Cemetery. Cricketer W. G. Grace is buried a few yards away.

Japanese. What is perhaps even stranger, is that we do not differentiate between the pan itself, the room where it is installed, and the building, at least in the case of public lavatories.

I did not really discover how unmentionable the subject is until I presented a short piece on television about Thomas Crapper, a Yorkshire lad who set up his own plumbing business in London in 1861 and was so successful that the firm still exists today (see www.thomas-crapper.com for details). Some people assert that Crapper invented the lavatory, or the flushing lavatory, or the siphon, but sadly none of those things is true. He did take out a few patents, but they were for things like ventilation of house drains and improved pipe joints. Thomas Crapper did not invent anything important. Nor did he give his name to 'crap', meaning 'excrement'. The word 'crap' comes from an old Dutch word meaning 'rubbish', and was used to mean excrement in 1845 – well before anyone could have heard of Thomas.

When the Crapper piece was televised, I was surprised at the interest shown by all sorts of people, and my partner, Sue, suggested I should write a book about lavatories. I thought this was a silly idea, but she persisted, and eventually I resorted to my training as a researcher and began to use my phone as a weapon, in order to ask everyone I could think of, from submariners to NASA, about their lavatories. The result was a book called *Thunder, Flush and Thomas Crapper*, which is now sadly out of print, and a lifelong interest in all things lavatorial. Unfortunately, the book is rather short, because my publisher told me to cut the crap, so I still have plenty of material waiting to be written. Here, however, is a hint of what we know of the history of lavatories.

All animals have to defecate to get rid of that unwanted waste material. Therefore human beings have always had to go. The early hominids probably went in the bush, but because we are programmed to hate the smell of faeces, they probably avoided going in the cave or wherever they lived. Only when people started farming and so settled in one place did it become necessary to make some sort of sanitary arrangements.

In 2004 I was delighted to have the opportunity to visit the oldest known lavatories in the world, at the little settlement of Skara Brae, on

the west coast of Orkney. In 1850 a great storm ripped the turf off some sand dunes, revealing nine or ten Stone Age houses. Huddled together, and built below ground level to escape the worst of the weather, these houses are all the same, with a fireplace in the centre of the single room, a big bed to the right of the door and a smaller bed to the left. In the corner is a small doorway that leads into a little cubicle perhaps 75 centimetres square, with a hole in the middle of the floor that goes down into a set of drains that interconnect and lead downhill towards the sea. These cubicles are en-suite lavatories and are five thousand years old. Seeing these myself was one of the highlights of my lavatorial career.

The oldest lavatory in the world, at Skara Brae on Orkney.

Other ancient lavatories have been found at Mohenjo-Daro, now in Pakistan, which are more than four thousand years old, and King Minos's palace at Knossos on the island of Crete also has the remains of a magnificent lavatory, perhaps two and a half thousand years old, with marble seats and a cistern for flushing. Alas, when I went there on a pilgrimage to see this sanitary marvel, the building was temporarily closed because of earthquake damage. Apparently, it had been temporarily closed for ten years...

The Romans, whose empire flourished around two thousand years ago, built magnificent public lavatories all around the Mediterranean, and even as far north as Hadrian's Wall. This northern boundary of their empire was a defensive rampart 75 miles long, built around AD 122 on the orders of the Emperor Hadrian; it went right across the north of England. The eastern end lies by the relatively modern city of Newcastle.

The latrine at the Roman fort at Housesteads on Hadrian's Wall.

I have visited Hadrian's Wall several times and am especially fond of Housesteads, one of the forts, where 800 soldiers lived while they patrolled the wall. In the south-east corner of the fort is the latrine, a room about 10 metres by 5 metres, which must have accommodated up to twenty men sitting shoulder to shoulder on seats over the sewer, which was continuously flushed with water from a cistern up the hill. The sewage ran out through a hole in the wall into the civilian settlement outside. The Romans seemed to have wiped themselves clean with sponges on sticks, and it seems likely that each soldier carried his own sponge, since you would not want to use someone else's, would you?

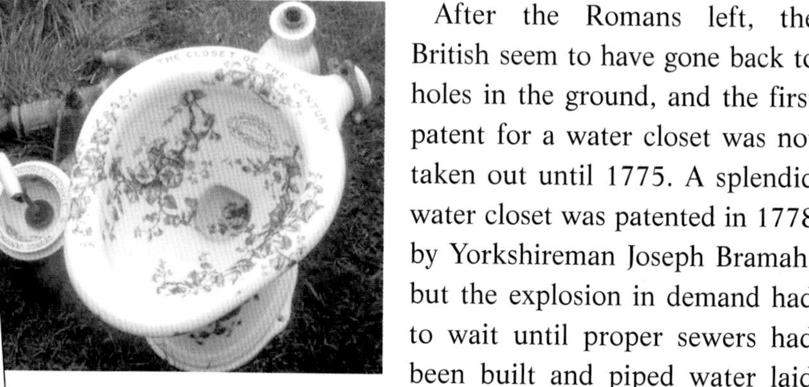

George Jennings took out patents in 1852 and 1854 for wash-out and wash-down lavatories like those in common use today.

After the Romans left, the British seem to have gone back to holes in the ground, and the first patent for a water closet was not taken out until 1775. A splendid water closet was patented in 1778 by Yorkshireman Joseph Bramah, but the explosion in demand had to wait until proper sewers had been built and piped water laid on, which did not really happen until the mid-1800s. The Great Exhibition of 1851 was staged in Crystal Palace, a vast glass-and-iron structure specially built in Hyde Park, where flamboyant plumber George Jennings installed public lavatories, and of the six million people who visited the exhibition, 827,000 chose to 'spend a penny' using his cubicles, which may well be where the expression comes from.

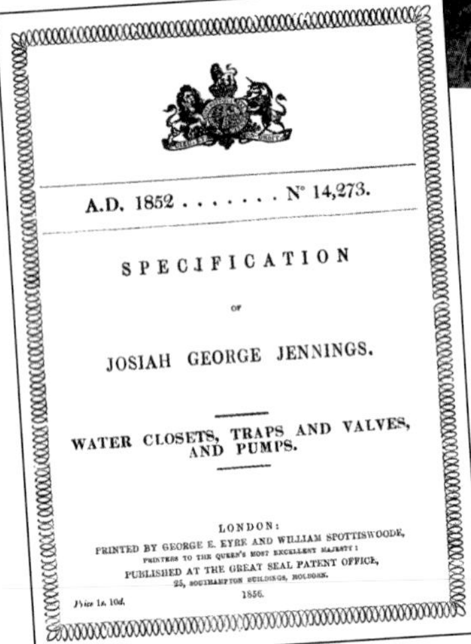

A.D. 1852 . . . . N° 14,273.

SPECIFICATION

OF

JOSIAH GEORGE JENNINGS.

WATER CLOSETS, TRAPS AND VALVES, AND PUMPS.

LONDON:
PRINTED BY GEORGE E. EYRE AND WILLIAM SPOTTISWOODE,
PRINTERS TO THE QUEEN'S MOST EXCELLENT MAJESTY:
PUBLISHED AT THE GREAT SEAL PATENT OFFICE,
25, SOUTHAMPTON BUILDINGS, HOLBORN.
1856.

Price 1s. 10d.

During the 1850s hundreds of patents were taken out for various forms of water closet, including flush-out and flush-down systems (the commonest types in use in the West today), loos that flushed when you pressed a foot pedal, and others that flushed automatically when you stood up. By about 1860 the basic patterns had been set, and Thomas Crapper was just too late entering the field, although no doubt he exploited the demand with success.

There was, however, one loud and dissident voice – that of the Reverend Henry Moule (pronounced 'Mole'), vicar of Fordington, just outside Dorchester, who championed the earth closet. He designed and patented his own earth closet in 1860, and sold many thousands of them: 148 were used by 2,000 men at the Volunteer encampment in Wimbledon in 1868, 'without the slightest annoyance to sight or smell'; 776 closets went to Wakefield Prison; Lancaster Grammar School brought in earth closets because the water closets were always out of order 'by reason of marbles, Latin grammar covers and other properties being thrown down them'. In a string of blistering pamphlets Moule declared that the water closet was an abomination, and merely shifted the decomposing sewage downstream. He said we should all take charge of our own effluent; then not only would we save water and

The Rev Henry Moule came to believe that the earth closet was much more hygienic than the water closet.

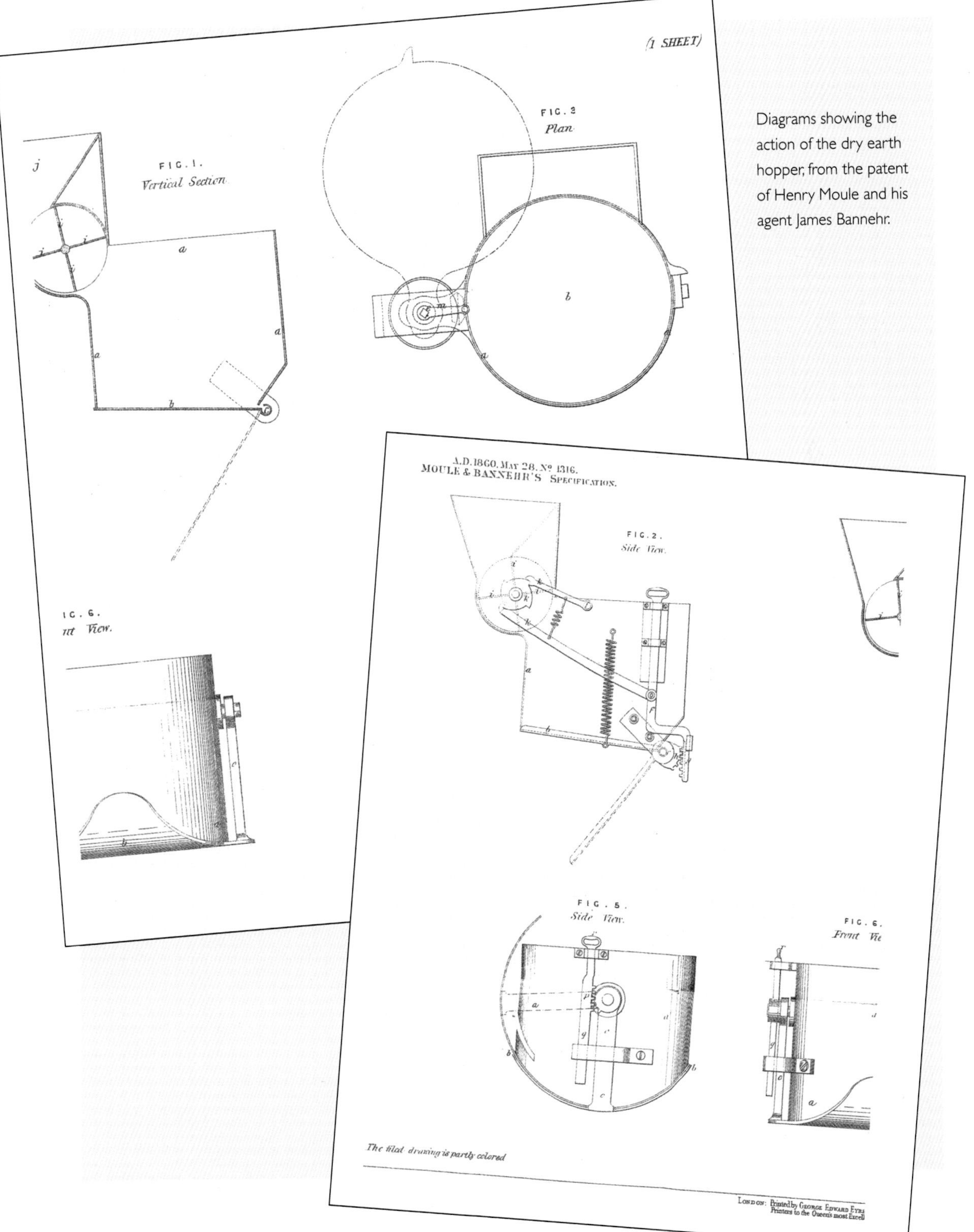

Diagrams showing the action of the dry earth hopper, from the patent of Henry Moule and his agent James Bannehr.

A Moule patent earth closet, in the Dorset County Museum in Dorchester.

money, but we would have a luxuriant growth of vegetables in our garden.

Similar arguments are used today by those in favour of composting lavatories. The environmental considerations have not changed: every time you use a water closet, you not only throw away valuable fertiliser, but you casually throw after it a couple of gallons of drinking water. What a waste! And flushing only takes the problem downstream; the sewage has to decompose somewhere. Basically, Henry Moule was right.

I have followed Moule's advice and made my own earth closet – essentially a bucket under a wooden seat with a hole in it. I collected earth from the garden, dried it in a pan over the boiler in the kitchen and kept it in a container beside the closet. The whole family used it

Between the wheel-barrow and the compost heap is my straw-bale urinal.

for a month during the summer, depositing a small trowelful of earth on top of each offering. No mess, no smell, it worked fine as long as we did not get it too wet by peeing in it, and we did indeed get wonderful vegetables later on.

Should you choose to try an earth closet, beware of one thing: intestinal worms. All early people were infested with them, and one major advantage of the water closet is that it breaks the life cycle of these worms. If you use an earth closet, leave the used earth to decompose for a couple of years before you reuse it, to make sure that the worms and their eggs have all died. A simple day-to-day problem, I discovered, is collecting and drying the earth, which is tedious; flushing everything away is much easier…

Whatever sort of lavatory I sit on, I always wash my hands afterwards, using plenty of water and soap if possible. Lavatory paper is a great advance on leaves and sponges, but it will allow your hands to be contaminated with bacteria and parasite eggs. Removing these is vital before you eat anything.

After peeing, washing is less important, since urine is generally sterile. We have a straw-bale urinal in the garden, permanently available for use by men, and any women who don't mind a little prickling. This is simply a small bale of straw stuffed into a plastic box with the cut ends upwards. You pee into it, the urine is absorbed by the straw and gradually decomposes inside the bale. Just like the earth closet, there is no smell, no mess, and no problem as long as it does not fill with rainwater; I have now built a roof to keep it dry. The straw bale lasts for perhaps six months, depending on how many people use it, then goes on to the compost heap and is later dug into the vegetable beds. As a result, we save valuable nitrogen fertiliser and many litres of expensive drinking water – about seven litres every time we don't flush the loo.

But I digress. After the lavatorial excursion, the time has come to set off for work.

4 Going to Work

HOME AND AWAY

I am lucky. For the last ten years I have had no job, which means that I have no meetings, no boss to shout at me and no commuting. The main disadvantage of being self-employed is that I get paid only when I actually do something that someone else likes, rather than for sitting at a desk for a specified period of time. The advantages are that I can work when I want, for as long as I want, and on what I want; if I choose to take photographs all day and read my emails at midnight, no one will complain.

Also, I make my own decisions, usually instantly; I do not have to summon a meeting and hear everyone else's views on what to do. Although I miss the constructive arguments that I believe are essential in the planning of any good television programme, I am glad to be spared the ideas of some of the idiots who seem to inhabit every office.

On good days I work at home, which means that going to work takes about twenty seconds – either downstairs from the bedroom or upstairs after breakfast. The regime varies, but I like to get in a bit of work – half an hour or even an hour – before breakfast, because at that time of day there is not much email and the phone does not ring; everything is peaceful, and my brain is in top gear.

Some weeks, however, I seem almost never to be at home; I travel elsewhere to film, record radio programmes, give talks and meet producers or publishers, or perhaps purveyors of charm, wit and useful information. For local travel, I use my bike, if I can. For greater distances, I take the train. Occasionally, I borrow my partner's car – I have no car of my own.

I owned cars for more than thirty years – for one glorious year I had an E-type Jaguar – but about ten years ago I discovered that I no longer enjoyed driving because the roads were increasingly congested (since 1980 the number of vehicles on the UK roads has doubled). I also found that I was using my car less than once a week, which meant that it was costing me about £50 every time I took it out of the garage – and that was not counting the soaring costs of fuel and maintenance. Since a taxi ride in my part of the world rarely costs more than £10, the car was clearly a terrible waste of money, so I sold it and now have a small cycle fleet poised for excursions.

CYCLING ADVENTURES

Cycling has always been a part of my life. I enjoy it and appreciate its scientific virtues. First, it is wonderfully low on energy consumption. According to the *Scientific American*, a person on a bicycle is the most efficient form of transport known, better even than a salmon. Apparently, 98 per cent of the effort exerted on the pedals goes into pushing you forward, whereas cars are only about 25 per cent efficient. To put it

Dressed for *Local Heroes*, on my pink-and-yellow bicycle, Jalopy (Just A Lovely Old Pink and Yellow).

Fig. 70. *Spinaway Tricycle.*

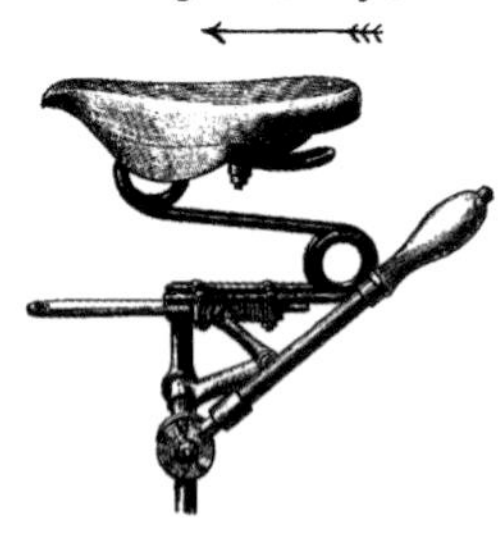
Fig. 71. *Sliding Seat.*

another way, a car needs to consume at least thirty times as much power as a cyclist, in order to go perhaps five times as fast. On short urban journeys the bike is usually quicker than the car.

Second, cycling is environmentally friendly. It runs on breakfast, produces no pollution or noise and scarcely affects the surface, especially on roads, whereas cars run on fossil fuels, fill the atmosphere with carbon dioxide and oxides of nitrogen, make a lot of noise and wear down the roads.

Third, cycling provides the rider with good exercise; no impact stress to hips or ankles, but a good cardiovascular workout. Cars provide no exercise, except when you have to push them.

The safety bicycle appeared around 1880 and has not changed much since. In the early decades, there were tremendous efforts to improve it in every possible way. I looked through all the British patents for ten days of the year 1897 and found that almost one in ten was for some improvement in cycle design. There were saddles with rollers in front so that ladies with long dresses could mount without fear of getting their dresses snagged, inflatable saddles 'to protect the organic parts', and brushes that would automatically clean the mud from rims and spokes. One new design used the surplus energy from coasting downhill to compress air and pump it into the hollow frame, then used the stored energy of the compressed air to push you up the other side, and there was even one bike made entirely from cane tied together with string. None of these things has lasted.

As in Darwinian evolution, only the fittest ideas survive. This is an example of memetic evolution, where memes are units of culture, in this case ideas for improvements in cycle design. Over the past century there must have been thousands of attempted improvements, but few of them have survived in the jungle of cycle memetics.

We now have some major variants, however. For one thing, we use better steel and other high-tech materials; we have better brakes, better tyres and many more gears. The fact that bikes now routinely have eighteen or twenty-one gears, whereas when I was a lad, three gears were the limit, has happened partly because it is technically possible, using modern materials, and partly because we are softies and don't like having to pedal too hard. Legs usually perform two functions: they help us to stand up and they push us forward. The bike cunningly gets rid of the first function, so all a cyclist's legs have to do is push forward.

In addition, we have some minor variations on the original safety bikes – the small-wheeled bikes, folding bikes and recumbents. Small wheels make a bike a bit smaller and highly manoeuvrable. They are useful on folding bikes to reduce the size of the folded package. A folding bike has the great advantage that you can take it in with you to most buildings – shops, libraries, your home – so that you don't have to leave it outside, where it can be rained on or stolen. In addition, you can take it in the boot of your car, or on the train, and use it at both ends of the major journey.

The Brompton folding bike – a brilliant design by Andrew Ritchie. I can fold it in 15 seconds and carry it anywhere.

I was born and raised on a farm in Oxfordshire, and the house was a mile from any public road, so riding a bike was fun and part of my childhood. I used to take it to pieces every now and then, and try to collect all the bits of the ball bearings that sprang out all over the floor of the shed; this was a vital part of my early training in engineering. I also needed the bike if ever I wanted to visit my friends. Later, as a student, I used a bike to get around Oxford; I had one of the early Moulton bikes with small wheels and simple suspension.

The tipping point came on 7 August 1990, while I was a producer at Yorkshire Television in Leeds. In an effort to keep fit and thin, I used to play squash at lunchtimes (when I was not playing chess), but fat old men die on the squash court, and I thought perhaps I should try something a bit less violent. So I bought one of those new-fangled mountain bikes – a British Eagle Trail – planning to ride into work two or three times a week, which meant a 25-mile round trip over three fierce Yorkshire hills. I deliberately chose a pink bike and put yellow mudguards on, in order to be visible, since I wanted to stay alive. I also bought a pink-and-yellow cycling jacket and a pink helmet.

I did manage that ride a fair number of times, and once I did it four times in a week, which left me feeling tired but virtuous. Although sometimes in January, when the ride home was all in the dark and the rain, and involved one 4-mile stretch of continuous uphill, I did question my sanity. What's more, all that cycling utterly failed to make me thin. A few years ago I rode 350 miles in a week and came home the same weight as when I left, probably because the exercise made me hungry and I ate more food than usual. The weight I lost by riding 50 miles a day was probably regained when I ate an extra slice of bread and butter, or drank an extra pint of beer in the pub that night. So by riding my bike I failed to change shape, but I did change direction, for I became a television presenter.

Joseph Priestley's portrait on the pub sign of what used to be his teenage home in Heckmondwike.

One day in the weeks after I bought the bike, I was trying a new route – I was always looking for the least bad way up the hills – and just before the bridge over the M62 on the long hill from Birstall to Drighlington, I crawled past a farmhouse, Fieldhead Farm, which had a blue plaque on the wall. Any excuse to stop for a rest. I discovered that Joseph Priestley, discoverer of oxygen, had been born there. I later found out his mum had died when he was nine, and he spent his teenage years with his aunt at the Old Hall in Heckmondwike, which had become my local pub, serving Sam Smith's beer and excellent sandwiches. Later on, Priestley discovered oxygen as a direct result of watching the beer fermenting in a brewery in Leeds. So here was a story of the discovery of oxygen, with a farmhouse, a pub and a brewery, all on one bike ride. This was the birth of my series *Local Heroes*, which ran for two years on Yorkshire Television and for six more years on BBC2. In these programmes, I rode around on my pink-and-yellow mountain bike talking about dead scientists and demonstrating what they did, in the places where they lived and worked. I wish we could make some more of these programmes since they were much fun and much loved, but sadly the BBC has other ideas.

At home, I ride either that old original mountain bike, Jalopy (Just a Lovely Old Pink and Yellow), which has had new paint, wheels, pedals, gears, brakes, saddle and mudguards, but is still the same bike at heart, or a wonderful three-wheeled recumbent trike (a Trice Explorer), lying back in style with my feet stuck out in front. This is hard work uphill, but superbly comfortable, with a soft webbing seat and no weight on the hands and wrists. I use the trike to go shopping for the family; I tow a lightweight trailer that will carry provisions for four for a week, plus cat food (see picture on the next page).

Setting off on the bike that changed my life for a 350-mile ride from Bristol to Darlington.

Carrying a week's shopping in the trailer behind my Trice Explorer.

Just a few times I have come to grief from bikes. I crashed bloodily when I was showing off at about the age of eight and swerved too sharply on a gravel road. I was knocked off once by a man who opened his car door in front of me while I was riding gently along a busy urban street. I have fallen off a few more times, once spectacularly while being filmed in Bath. But so far I have suffered no serious injury.

My most terrifying cycle ride was in a game reserve in south India in 1961. Between Mysore and Ootacamund, I got off the bus at Mudumalai and arranged with the range officer in the tiny village to ride into the jungle on an elephant the following morning. He told me that in that block of jungle – 10 square miles – there were ten tigers, twelve panthers and a rogue elephant, and we might see any of them. He told me to stay in a guest house two miles along the road and come back to the village at 5 a.m. The walk to the guest house was dramatic: on one side of the road there was a 50-foot drop into a stream, and on the other solid jungle came right to the edge of the road; the air was filled with the screeches of birds and monkeys.

I had a pleasant evening at the guest house, writing postcards home, and the proprietor woke me at 4 a.m. with some tea. I asked how I could get back to the village and he offered to lend me his bicycle. So I set off along the road. The bike had no lights, but I was not worried: there is always a little light when your eyes grow accustomed...only that night there wasn't. I held up my hand in front of my face, but could not see it.

I had two miles to go, and I could think of nothing but the facts that along the right side of the road was a 50-foot drop into a stream, while on the left was solid jungle, containing ten tigers, twelve panthers and a rogue elephant. Had one of these beasts been standing in the road, I would have gone straight into it. I went slowly, keeping my left foot on the grass verge so that I could not go off the right-hand side. I did not know whether to be as quiet as I could or whether it would be better to sing loudly. The trip to the village seemed a long way indeed.

WHAT TO WEAR ON THE WAY

If the weather is hot, I would like to stay cool when cycling, so I wear cycling shorts and a T-shirt, but when it may be cold for at least some of the time, then I need to choose clothes with better heat insulation. For a complicated journey, I may keep warm cycling, get cold standing on the platform waiting for a train and then warm up again in the heated train, so I need to have some way of varying the insulation – either layers that I can put on and take off as necessary or a jacket with a front that I can zip right up in the cold or leave partly or completely open. The important thing is to have enough insulation to cope with the lowest temperatures.

PLANNING YOUR OUTFIT

Choosing the right clothes for travelling poses five separate questions, which is why I find it so easy to choose wrongly.

1 What sort of temperatures am I likely to encounter?
2 Am I going to sweat?
3 Is it raining, or likely to rain?
4 What is the wind speed?
5 What sort of fashion statement do I want to make?

While riding a bike, I generate a good deal of heat, so I rarely get seriously cold as long as I have a windproof layer on top, although I do need gloves when the temperature is below freezing. Cold becomes a serious problem when you are not moving, however. In western Canada, I have experienced temperatures as low as −40°C (which is also −40°F), but I was ready for that and took sensible precautions. We always wore gloves outside; we had thick coats; we plugged our car into the socket provided at every parking space in order to keep the oil from solidifying. I got used to driving off with 'square wheels', when the car tyres had frozen flat underneath, to the prickling sensation of icicles growing down from my nose when I breathed out, and to my spectacles frosting over when I went from the cold into a warm, damp building.

Wearing gloves is important in extreme cold, since most surfaces have a dusting of snow in the winter. Take hold of, say, a doorknob with your bare hand, the warmth melts the snow and then your hand freezes to the doorknob. Most unpleasant. One foolish lad licked an iron railing on the way home from the movies and his tongue froze to the iron. Fortunately, his dad had the presence of mind to pee on it; the warm urine provided enough heat to melt the ice for long enough for the boy to remove his tongue. Think about it: would you rather have a mouthful of urine, or spend the night attached to a railing by your tongue?

Snowflakes are crystals of ice and usually have hexagonal shapes.

If I realise the temperature is going to be low, I take the necessary precautions, but I have felt perishingly cold on various occasions when I went filming without being properly prepared. Perhaps the worst time was one Easter night, filming pop-up urinals in Apeldoorn, east of Amsterdam. These new devices were the first in Europe, and the town council hoped they would prevent men from staggering out of the bars late at night and peeing on any convenient building. I had left home on a balmy spring day – we had coffee in the garden in warm sunshine – and I foolishly thought that the weather would be equally mild in the Netherlands. By midnight, however, it was well below freezing and hailing with venom, and I was horribly, horribly cold. I keep forgetting how cold it can be on location. Often I feel that the temperature drops sharply within 100 metres of any film crew – and rain often begins to fall in the same area.

Travelling abroad, I often make the same mistake. Not being sure what the weather will be like, I tend to take too many thin clothes and not enough thick ones. This is foolish, for taking clothes off – or buying thinner ones – is easy, but when I have only T-shirts and there is snow on the ground, I feel like an idiot. And yet I still do it. On one memorable occasion I flew from Albuquerque, where it was warm even in December, via Chicago to Toronto. I only just made the connection in Chicago, and my luggage didn't, so I arrived in Toronto in the evening wearing only trousers and a T-shirt, which I had then been wearing for two days. The next morning I had a breakfast meeting with a man who insisted on showing me the city, on foot, in the snow...

At the other end of the temperature scale, heat can also be a problem. I try to avoid cycling if the temperature is above 30°C, but even on a cool day I can generate a good deal of sweat. Sweating is important for cooling the body (see page 45); it is mainly water, and as the water evaporates, it extracts the latent heat from your skin. This is good if your skin is exposed to the air, but not so good if it is enclosed in a waterproof layer. Wicking is important and so is a breathable outer layer – one that will allow the sweat to evaporate, so that your clothes don't get soggy. The best material I know is Gore-tex, a remarkable fabric that is breathable but waterproof.

When I am going to be on the road for more than an hour or two, I generally assume that rain will fall; getting wet is unpleasant, but rainproof gear is not too heavy or awkward. Gore-tex is made of a tough synthetic fibre, plus a thin film of PTFE (polytetrafluoroethylene – the same stuff as on non-stick saucepans). This PTFE layer is perforated with millions of minute pores, each one thousands of times smaller than even the tiniest drop of water, but hundreds of times bigger than a molecule of water vapour. Liquid water is strongly repelled by the PTFE and cannot penetrate the pores, but water vapour can escape, so that your sweat does not build up inside and make you soggy. Gore-tex is not warm at all, but can be worn over warm inner layers, and my warmest coat has a Gore-tex outer layer and a thick fleece lining.

I believe in wearing bright colours when cycling in order to be visible to motorists.

PTFE was discovered on 6 April 1938 by a research chemist called Roy Plunkett who wanted to do some work with the gas tetrafluoroethylene, but when he opened the valve on the cylinder, no gas came out. The cylinder seemed heavy enough to be full, so he was surprised, and took the rather drastic and potentially dangerous step of sawing through the cylinder

Relentless rain during filming in Borrowdale: showing off a lump of wad (see page 113).

to find out what had happened. Inside, he found a white waxy solid: rust inside the cylinder had catalysed the polymerisation of tetrafluoroethylene to PTFE, which turned out to have some remarkable properties; it is resistant to heat and incredibly slippery; nothing will stick to it.

Gore-tex was also discovered almost by accident, in 1969, by Bob Gore, who was trying to stretch a hot PTFE rod and found he could expand the material to make a thin sheet.

When I am walking rather than cycling, I prefer to wear something other than a cycling jacket, and having recently had to invest in a coat, I thought I should investigate the history of waterproofing. The earliest waterproof garments we know about were made around 1600 BC by the Meso-Americans, who made useful rubber by treating raw rubber latex, or *caoutchouc*, with the sap of the morning-glory vine, but the subtle chemistry involved took some time to reach Britain.

THE DEVELOPMENT OF RUBBER

The English chemist Joseph Priestley was intrigued by *caoutchouc* and in 1770 found it was good for erasing pencil marks, so he suggested it should be called 'rubber'. In 1824 Michael Faraday discovered that by pressing the edges of two sheets of *caoutchouc* together, he could make containers for the hydrogen gas with which he was experimenting. 'The *caoutchouc* is exceedingly elastic,' he wrote in the *Quarterly Journal of Science*. 'Bags made of it have been expanded by having air forced into them, until the *caoutchouc* was quite transparent, and when expanded by hydrogen they were so light as to form balloons with considerable lifting power.' Toy rubber balloons were introduced in 1825 as messy-sounding kits – you had to make them yourself – and vulcanised rubber balloons were first made by J. G. Ingram of London in 1847.

Charles MacIntosh's patent for waterproofing fabric.

A.D. 1823 . . . . N° 4804.

Rendering Fabrics Waterproof.

MACINTOSH'S SPECIFICATION.

TO ALL TO WHOM THESE PRESENTS SHALL COME, I, CHARLES MACINTOSH, of Cross Basket, County of Lanark, Esquire, send greeting. **WHEREAS** I, the said Charles Macintosh, did, by my Petition, humbly represent unto His present most Excellent Majesty King George the Fourth that I had, after much study and expence, discovered and invented "A PROCESS AND MANUFACTURE WHEREBY THE TEXTURE OF HEMP, FLAX, WOOL, COTTON, AND SILK, AND ALSO LEATHER, PAPER, AND OTHER SUBSTANCES, MAY BE RENDERED IMPERVIOUS TO WATER AND AIR;" that I was the first and true Inventor of the said improvement, and that the same had not been practised or used by any other person or persons, to the best of my knowledge and belief, and praying that His Majesty would be graciously pleased to grant unto me, my exors, admors, and assigns, His Royal Letters Patent under the Great Seal of the United Kingdom of Great Britain and Ireland, for the sole use, benefit, and advantage of my said improvement, within that part of His Majesty's United Kingdom called England, His Dominion of Wales, and Town of Berwick-upon-Tweed, and also in all His Colonies and Plantations abroad, for the term of fourteen ... Statute in that case made and provided: And whereas

The first person to make successful waterproof clothes in Britain was a Glaswegian, Charles MacIntosh. His recipe was to dissolve about 10 oz (300g) grams of *caoutchouc* in a 'wine gallon' of coal oil, which was a waste product of coal-gas production, in order to make a rubber solution. He painted this on a sheet of heavy cotton fabric and fastened another sheet of cotton on top. He patented his invention in 1823, and the mackintosh was born. Old-fashioned mackintoshes had a funny smell, they were heavy, and they were not completely waterproof, but they were better than nothing. Modern materials can be completely waterproof, but then they collect sweat inside. That is why Gore-tex, waterproof and breathable, is so effective.

What is surprising is that MacIntosh managed to make his fabric waterproof using *caoutchouc,* which was nasty sticky stuff and difficult to handle; in hot weather it melted, while in cold weather it froze and cracked. Rubber as we know it was first made by mistake in 1839 or 1840 by an American hardware merchant called Charles Goodyear. He had been experimenting with all sorts of materials in an attempt to tame raw rubber, and while heating it with sulphur (and white lead), he spilled some of the mixture on the hotplate. The smell was disgusting, but the result was useful tough rubber. What he had done was to cross-link the long hydrocarbon chains of the natural rubber with sulphur atoms; he called the process 'vulcanisation', after Vulcan, the Roman god of fire. Once rubber had been vulcanised, it became extremely useful, and exports of raw material from Brazil increased from 31 tons in 1827 to 28,000 tons in 1900.

My umbrella hat: do I look silly in this?

UMBRELLAS AND HATS

A simple solution to urban rain is the umbrella; I have one that is big enough for three people, which I take when heavy rain is already falling as I leave the house, and a small folding umbrella that I stow away in my briefcase when rain appears to be imminent. I even have a tiny one that I can wear like a hat, although it seems to me I may as well wear a hat. The advantage of the umbrella is that the shape makes the rain run off quickly, and it never soaks through, even when the material is not waterproof. I find umbrellas are effective at keeping the rain off my head and shoulders, but they tend to let it in lower down and they are awkward in a crowded street, always in danger of getting caught by the wind or on lamp posts, and of poking someone else in the eye.

The mathematics of running in the rain are surprisingly complex, but luckily an experiment has been done by Thomas Peterson and Trevor

TO RUN OR NOT TO RUN...

When the rain starts and I have no umbrella, I sometimes wonder whether I should run. This turns out to be a complicated question, and the answer depends on several factors:

1 If I am wearing a waterproof coat with a hood, and I don't care too much about my trousers, then there is not much point in running, but if I am wearing smart clothes, then I would like to minimise my exposure to the rain.

2 If there is no wind and the rain is falling vertically, there is an argument for walking slowly, so that the rain falls only on my head and shoulders. If I run, I will sweep out extra rain with my body – or in other words, because I am moving forwards, I will collect with my body raindrops that would otherwise have fallen harmlessly in front of me. What is more, the faster I run, the more raindrops I will intercept every second. On the other hand, the faster I run, the quicker I will reach shelter, and so the less time I will be exposed to the rain.

3 If the wind is blowing and the rain is falling at an angle rather than vertically, then I am definitely going to get wetter, since the rain will hit not only my head and shoulders but one side of my body too. In this case, running is definitely a good plan, if only to shorten the time of exposure. If the rain is coming from in front or from the side, I try to lean into it in order to reduce the surface area exposed to the rain. If the rain is coming from behind, then I do my best to run at the same speed as the wind, since if I could achieve this, then I should be moving forward at the same speed as the raindrops. As a result, they would fall only on my head and shoulders; this would be like standing still in vertical rain.

Wallis of the US National Climatic Data Center. They put on identical sweat suits and hats and made a 100-metre journey in the rain, one running, the other walking. Then they weighed the clothes to see which had become wetter. The runner had collected 138 millilitres of water, the walker 231 millilitres (see www.straightdope.com/classics/a3_395.html for details). So it is worth running in the rain.

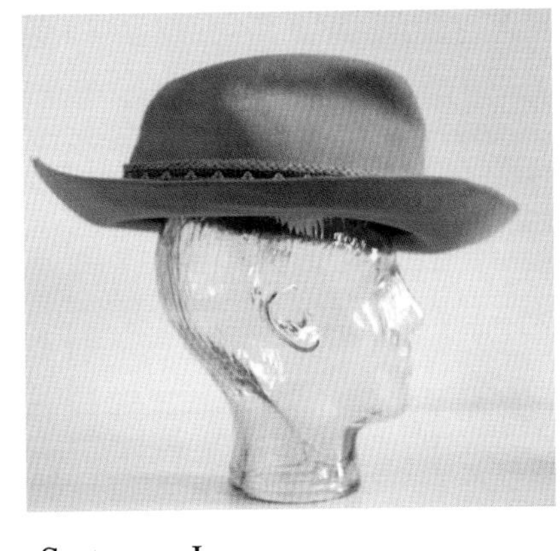

However, I dislike running, so I always try to remember to carry my folding umbrella, or else I wear a hat. I have a fine collection of hats, even though a number have been lost over the years, including my pith helmet, my fez and two Stetsons. I still have a battered brown Stetson, bought in Houston, Texas, which is good at protection from rain, although perhaps not quite as good as my Australian leather bush hat. Some Australian hats have corks hanging from the brim on strings to keep the flies away, but mine is cork-free. When I set off for a walk in the hills and the weather is doubtful, then I

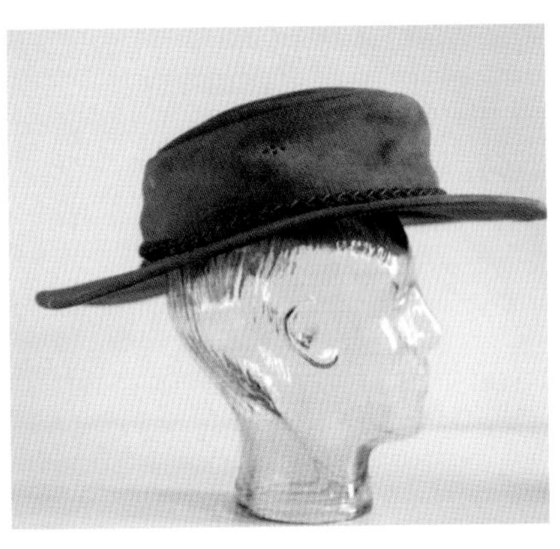

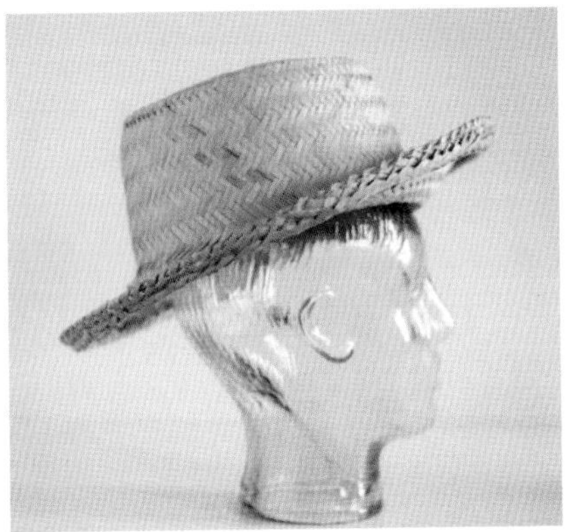

A selection of my hats: Stetson (above), corkless Aussie bush hat (far left), African straw hat bought at the market in Kisumu (left) and floppy cotton hat (below).

may decide to rely on the hood of my waterproof jacket, which is truly waterproof but restricts my hearing and my peripheral vision, or I take my Aussie bush hat, which keeps the rain from my face even better than the hood and does not interfere with my hearing or vision.

Some hats I wear for warmth – my soft trilby or fedora, my woollen hat from Kashmir, my silly knitted hat from the French Alps and my fleecy hat for filming. I have an African straw hat that kept the sun out of my eyes in Kenya, and a floppy cotton hat that did the same job in Greece. I have a

Battered stovepipe hat in the style of I. K. Brunel, who built the SS *Great Britain*, behind me.

baseball cap that I wear under a cycle helmet to keep rain off my spectacles. And I have a couple of hats that I wear purely as fashion statements. The most spectacular is an enormous battered stovepipe hat of the kind worn by Isambard Kingdom Brunel. His 200th birthday is being celebrated in 2006, and I bought this fine hat to wear during the celebrations, which are many and various.

Finally, I have a Coke hat, of which I am proud. When Victoria came to the throne of England in 1837, she created the First Earl of Leicester, and his nephew William Coke (pronounced 'Cook') was worried about the gamekeepers on the family estate, Holkham Hall, in Norfolk. He reckoned they were in danger of attack from poachers and low branches, and to protect their heads, he designed a tough hat – a rounded felt hat, stiffened with shellac, with a semi-rough finish. He went to the family hatters in London, James Lock & Co. of St James's, and asked whether they could make him a prototype. A week later he returned, tried it on – it fitted – and then put it on the floor and jumped on it. The hat did not collapse. The hatters may have been somewhat startled to have this scientific experiment conducted in their hallowed premises, but Mr Coke was delighted; he ordered a dozen, and the Holkham Hall gamekeepers started wearing them.

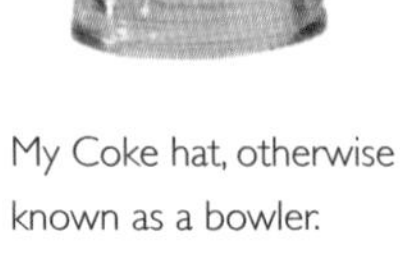

My Coke hat, otherwise known as a bowler.

Because the hat was originally made for Mr Coke, Locks still call it a Coke hat, and it is also known as a billycock hat, after its designer, William Coke. However, Locks did not actually make the hat themselves; instead, they put the job out to the brothers Thomas and William Bowler in Southwark. So to the rest of the world this has become the bowler hat. The gamekeepers at Holkham Hall still wear such toughened hats today, but when fashion dictated that all city gents should wear bowlers in the middle of the twentieth century, they became lighter and more of a fashion item.

One slight problem I face is that when I ride a bike I wear a cycle helmet, which prevents me from wearing any other hat – apart from the baseball cap in the rain – so the hats don't get as much use as perhaps they should.

Cycle helmets are made of foamed polystyrene, which is perhaps 20 millimetres thick, and I hope that if I come off my bike and land on my head on the road, the polystyrene will squish and crumple and protect my head, just as William Coke hoped that the tough hats he designed would protect his gamekeepers. It should also provide protection if my head is scraped along the ground. Sometimes on location in, say, building sites, I have to wear a hard hat; these have a tough outer shell held about 30 millimetres away from the skull by internal webbing, so that if a spanner falls on my head, the outer shell can deform this much before it hits my head.

Human beings lose body heat all the time, especially in cold weather, and about 20 per cent is lost through the head, so some people claim that wearing a hat is necessary to retain heat. I suspect that much of that heat is lost by breathing out, so although I do wear hats for safety and to keep off rain or fierce sunshine, I rarely do so for warmth – more often mine are for style and elegance.

Coke hats were designed for the protection of the gamekeepers at Holkham Hall, and are still worn there today.

I needed a hard hat and protective clothing down this sewer in Belfast.

LONDON TRANSPORT

Getting about in cities has always been a problem, and always will be, but in some ways London is unusually difficult. Before Victorian times there weren't many people living there, but by 1850 the population had grown to more than two million, and daily commuters swelled the numbers further. The streets were stinking with manure and crammed with pedestrians, people on horseback, horse-drawn carts and carriages – a ghastly crush. An omnibus service was tried, but failed because of the traffic jams.

The main-line railways were making things worse. The first terminus was London Bridge in 1836, where commuters arrived from Greenwich, and was followed in succession by Waterloo, Victoria, Paddington, Euston and King's Cross (1852). A board of commissioners had decided that driving main-line railways into the heart of the city would cause intolerable disruption to traffic and life, and as a result, each railway company had to build its own terminus; those north of the centre were all built along what is now the Marylebone Road and Euston Road.

Apart from Paris, which has several main-line stations, London is most unusual. Major cities in continental Europe generally have a central main railway station (*Hauptbahnhof* in German), from which lines go off in all directions. London, by contrast, has a ring of twelve main-line termini, and the Victorians had to find some way of getting people across and around this ring.

Successful trial run in the Metropolitan Railway, 6 September 1862.

The world's first underground railway was the Metropolitan Line, which went originally from Bishop's Road, Paddington, to Farringdon, and was opened on 9 January 1863. Before it was built, *The Times* described the idea as 'an insult to common sense'. But after the successful opening, when 30,000 people travelled along the line on its first day, *The Times* called it 'a great success' and 'the great engineering triumph of the day'. One of the few grandees who did not ride on that day was the seventy-nine-year-old Prime Minister, Lord Palmerston, who said he hoped to remain above ground a little longer.

Initially, the underground trains were pulled by steam locomotives, and the smoke and steam in the tunnels must have been chokingly awful, although the Metropolitan and other early lines ran for most of the time in cuttings, rather than deep underground, so that the smoke and steam had plenty of chances to escape. The first deep 'tube' line was the City and South London Railway, which ran from King William Street in the City to Stockwell. It was opened in 1890 and was the first underground railway to use electric power rather than steam.

Being the pioneer is not always a good thing. Passenger railways were invented in Britain, and because the decision was taken not to build them into the centre of London, they had to be supplemented by underground railways. As these date from Victorian times, they were not planned for the number of people who use them today. Expanding road transport capacity is simple – we can build more roads. Expanding surface rail transport is difficult and expensive. Expanding the underground is immensely difficult and expensive. We could make longer trains, but they would not fit into the station platforms, unless every platform were to be extended. We can't make wider trains without widening all the tunnels. The London Underground today is horribly crowded and most unpleasant in summer rush hours, but improving it is not an easy task.

One feature of the underground that is an enduring triumph is the map, designed in 1933 by a young unemployed draughtsman called Henry Beck. Abandoning the usual idea of a geographically accurate representation, he instead arranged the routes in straight lines and smooth curves, maintaining all the correct connections, but keeping the whole network tidy, compact and clear. When you travel by underground,

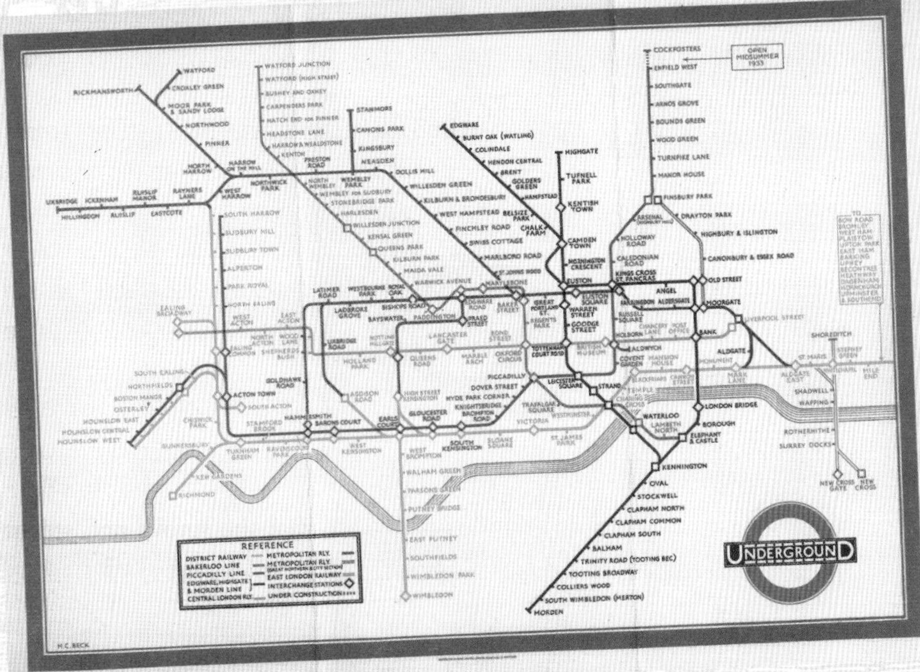

Henry Beck's original underground map.

you do not mind whether your train goes north, south, east or west, or how it wiggles on the way; all you need is the name of your destination and a simple diagram that shows you the quickest way to get there. That is exactly what Beck provided, and we should all be grateful to him.

When I have to go to London, which is sometimes only once a month, but for two or three days in some weeks, I take the train from Bristol, where I live, and then travel around the capital either by underground or more often by bike, which is fun, convenient and highly reliable: I am not stopped by traffic jams, by security alerts or by strikes. I used to take my bike with me on the train. My mountain bike had to go in the guard's van, which was a slight bore but never a serious problem, and I have not yet been turned away because the van was full. Then I bought a folding bike and began to enjoy an integrated transport system.

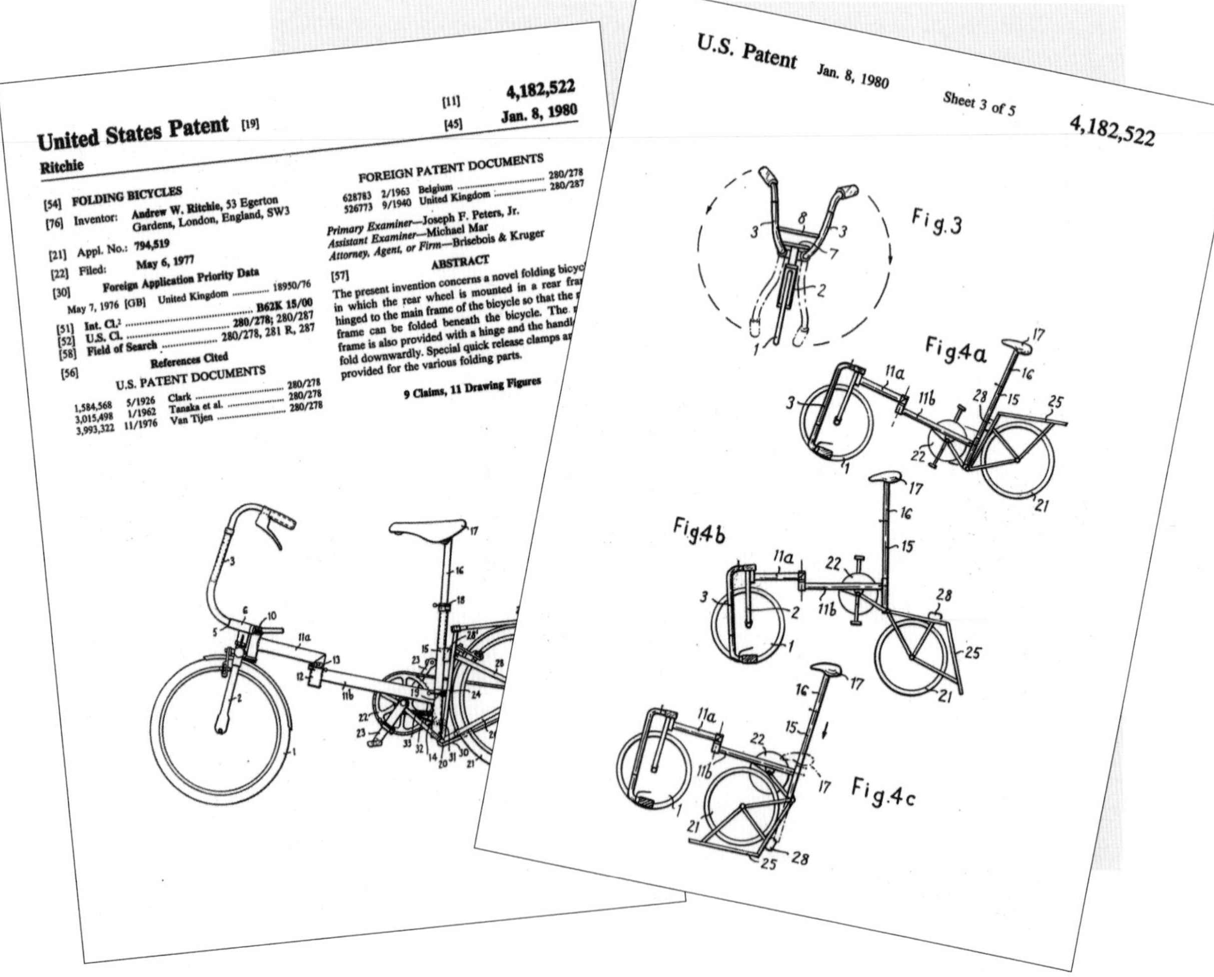

United States Patent [19]
Ritchie

[11] 4,182,522
[45] Jan. 8, 1980

[54] FOLDING BICYCLES
[76] Inventor: Andrew W. Ritchie, 53 Egerton Gardens, London, England, SW3
[21] Appl. No.: 794,519
[22] Filed: May 6, 1977
[30] Foreign Application Priority Data
May 7, 1976 [GB] United Kingdom 18950/76
[51] Int. Cl.2 .... B62K 15/00
[52] U.S. Cl. 280/278; 280/287
[58] Field of Search .... 280/278, 281 R, 287
[56] References Cited
U.S. PATENT DOCUMENTS
1,584,568 5/1926 Clark 280/278
3,015,498 1/1962 Tanaka et al. 280/278
3,993,322 11/1976 Van Tijen 280/278

FOREIGN PATENT DOCUMENTS
628783 2/1963 Belgium 280/278
526773 9/1940 United Kingdom 280/287

Primary Examiner—Joseph F. Peters, Jr.
Assistant Examiner—Michael Mar
Attorney, Agent, or Firm—Brisebois & Kruger

[57] ABSTRACT
The present invention concerns a novel folding bicyc in which the rear wheel is mounted in a rear fra hinged to the main frame of the bicycle so that the r frame can be folded beneath the bicycle. The frame is also provided with a hinge and the handl fold downwardly. Special quick release clamps ar provided for the various folding parts.

9 Claims, 11 Drawing Figures

U.S. Patent Jan. 8, 1980 Sheet 3 of 5 4,182,522

Andrew Ritchie's patent for the Brompton folding bike.

My London transport today is this folding bike – a pink-and-yellow Brompton L5 – which has enough gears to get me up the serious hill to Hampstead Heath and a big enough bag on the front to hold a day's work. At my destination, I can either chain it to a lamp post or fold it up and take it inside with me. The bike has been banned from the British Museum but is welcome in most establishments, including the Royal Society and the Royal Institution. The Brompton is not particularly light to carry, nor particularly beautiful, but it is a stunningly clever design. I can fold it in less than fifteen seconds into a bundle small enough to fit between the seats in a train, or on the luggage rack at the end of the coach, and when I unfold it, the gears are ready to go, the chain in place. If the heavens open while I am riding in London, I can carry it on to an Underground train, or take it with me in a taxi.

From Paddington Station to the BBC at White City, which is about 3 miles due west as the crow flies, takes me just thirty minutes on the bike, going at a steady pace through quiet back streets, and only for a few hundred metres is there any serious traffic. Underneath a vast concrete roundabout on Westway, I ride past tennis courts and a riding school, with small ponies plodding round under small girls. Only if I have something large to carry or if heavy rain is falling do I use the Underground, or taxis, and I have not yet fathomed the London bus system.

THE DEVELOPMENT OF THE WHEEL

All these various modes of transport use wheels, and many people hail the wheel as the greatest invention of all time. It probably originated some five thousand years ago in Mesopotamia, where Iraq is today, and developed from the idea of using logs as rollers to move heavy loads about. The first attempt at making a wheel may have been to cut a slice from a log and put an axle through the middle, but that does not work well because the centre part of a tree is weak and the axle would quickly have crushed and destroyed the wood around it.

If they had had huge trees, those Sumerians could have cut a circular slice that did not include the centre, but their trees were not big enough, so the next clear stage was a wheel made from three planks fixed together side by side, with the axle through the middle plank. This worked reasonably well, but was liable to break on corners,

Putting the red-hot iron tyre on a cartwheel.

because cornering a chariot or cart puts sideways stresses on the wheel, and the three planks could easily buckle along the seams when pushed sideways.

Next came the spoked wheel, possibly invented by the Greeks. This represents a significant step forward in technology. The hub, nowadays turned in ash or elm, was thick – chunky and strong to withstand the forces from the axle. Sunk into it were a dozen or more tough oak spokes, of which the outer ends were socketed into two or three bent ash timbers or 'felloes' that formed the inner rim. The outer rim, or tyre, came to be made of iron, heated to red hot, dropped over the wooden rim and rapidly cooled with water. As the iron cooled, it contracted, shrinking tightly on to the wood and forcing all the spokes deep into their sockets.

The spoked wheel is strong in all directions, can withstand those sideways stresses on cornering and has lasted for hundreds of years. Indeed, it is still used not only for fancy carriages but also for cars and trucks and trains, although their rims, spokes and hubs are made of steel. In these spoked wheels, the spokes are in compression, and the weight of the vehicle rests, through the spokes below the hub, on the rim and therefore on the road underneath.

The next major advance came almost incidentally, while Sir George Cayley was inventing the aeroplane in the first half of the nineteenth century. In the summer of 1853 he flew his coachman some 200 metres across Brompton Vale, behind his house in North Yorkshire. The aircraft was only a glider, but this was the first flight by a man in a heavier-than-air machine, and when the Wright Brothers flew their first aircraft fifty years later, they paid Cayley a handsome tribute.

Cayley wanted to make his aircraft as light as possible, and so redesigned the wheels to have spokes made of thin wire rather than thick timber. The hub of the wheel would then hang from the rim above, so that the spokes would all be in tension rather than compression. This is how cycle wheels work to this day, and because wire in tension is extremely strong, the spokes can be much thinner and lighter than spokes in compression.

There was one further major innovation – the pneumatic tyre. First invented by a Scot called Robert William Thomson in 1845, the air-filled tyres failed to catch on, even though Thomson laid on a demonstration for journalists in London's Regent's Park and showed that his 'aerial wheels' were quieter and more efficient than conventional wheels. The problems were that at the time rubber was difficult to obtain, and almost impossible to handle, since vulcanisation (see page 97) had not yet become a commercial reality. In any case, the bicycle and the motor car had not been invented, so there was little demand for effective wheels.

The pneumatic tyre was reinvented in 1889 by another Scot, John Boyd Dunlop. Dunlop was a vet, working in Belfast, and his son, Johnny, complained that riding his tricycle on the cobbled streets made his bottom sore. So his dad made some air-filled rubber tyres and fastened them to the wooden rims, the story goes, with strips torn from one of his wife's old dresses. Johnny soon found the tricycle was not only more comfortable but also faster than those of his friends. The pneumatic tyre makes travelling more efficient, as Robert Thomson had shown, because each bump in the road is absorbed at source by the flexing of the tyre and compression of the air in the tube, rather than throwing the entire wheel into the air with a scrunch.

Modern mountain-bike wheel, with gear block.

Realising there might be a future for his invention, Dunlop persuaded Willie Hume, captain of the Belfast Cruiser Cycle Club, to use pneumatic tyres in an important local cycle race at Queen's College Playing Fields on 18 May 1889. Everyone laughed at Hume when he turned up with these namby-pamby tyres, but they stopped laughing when he won the race. Then everyone wanted pneumatic tyres, and the Dunlop Rubber Company was formed. Vulcanised rubber was now commercially available, and Dunlop's tyres immediately became essential for all bicycles, and eventually for motor cars too.

The wheel has come a long way, and certainly supports most of our transport today, but it has not always been so successful. Indeed, the Ancient Egyptians actually abandoned the wheel for more than a thousand years. They had wheels three thousand years ago, but they stopped using them because Egypt is 95 per cent desert and the wheel was useless on sand; camels or sledges were more effective. Likewise in South and Central America, where the tracks between villages were narrow and steep, the wheel was not used in real life, although the Aztecs, Mayas and Incas had wheeled toys. Even in Devon and Cornwall, in the south-west of England, there were essentially no wheeled vehicles west of Exeter until the nineteenth century. The fact is that wheels are useless unless you have good smooth roads to roll them on.

USING COMMUTING TIME

Spending an hour or two travelling to and from work seems a terrible waste of time, and I should find it frustrating if I had to do it every day. But some people use the time they spend on trains and planes to work with papers or laptops, to read interesting or improving books, or even to write. Christopher Wood, who commuted every day in the late 1960s, decided to use his time on the train to write down some fantastical stories, using the pen name Timothy Lea. The result was the book *Confessions of a Window Cleaner*, and after a string of further *Confessions* bestsellers, he had made enough money to give up work; so he never had to commute again.

Writing on trains is difficult if you are using pen and paper, as some authors still do, because the unpredictable lurching makes it almost impossible to write legibly. I remember a splendid gadget like a wooden tray on which you put your sheet of paper. Then you strapped your elbow loosely to the corner of the tray. When the train lurched, the tray, resting on the table, went with the lurch, but so did your elbow and your writing hand, with the result that your handwriting was almost unaffected by the unpredictable movements. All that has now changed with the advent of laptop and palm computers (see page 126).

Other people's mobile phones have become a major distraction on trains. When they first became popular, everyone who had one seemed to spend most of the journey calling friends to shout, 'Nigel, NIGEL...I'm on the train – YES, ON THE TRAIN...' but now everyone has one, and many of them seem to be quite happy to share their intimate memories and fantasies with all those within earshot. Luckily, most train operators provide quiet carriages where the use of mobile phones is banned. That is where you will usually find me.

None of my books has been a spectacular bestseller, but I have written substantial chunks of them on trains, in aeroplanes and even in taxis. Curiously, although I am surrounded in those environments by noise, bustle, anxious parents and fractious children, I find I can switch off the outside world and focus on what I am trying to write. Sometimes I feel that the process of travelling enables me to write – a whole new meaning of 'going to work'. My latest New Year's resolution is to spend some time every day in the tiny summerhouse in the garden, escaping from the phone and the email, and trying to recreate that little haven of peace. Perhaps I should record some airport or train hubbub and play it in the summerhouse to complete the effect, so that I can actually believe I am working on the move.

At least half of this book was written on trains, using my iPAQ with a folding keyboard.

5 Working

ALWAYS SOMETHING TO DO

For me, work generally means one of the following:

-  Sitting in front of my computer at home answering email and ordinary mail.
-  Writing anything from an article for a newspaper to a chunk of the next book.
-  Doing research for a piece I am writing, using reference books, experts on the phone or in person, or the Web.
-  Taking photographs, sometimes out on location, but more often of scientific subjects in close-up, with all the apparatus fixed to a frame on one side of my room.
-  Recording pieces for a radio programme, either in a studio or on location.
-  Recording pieces for television, occasionally in a studio but more often on location, and often in scary places. When the producer discovers I am afraid of heights, for example, he or she usually sends me up the highest thing in sight...
-  Or giving a talk, often in the evening, to a large audience.

I enjoy all these activities, apart from the heights, and am especially lucky to have such a variety of things to do. What is more, they all feed off one another; I will pick up a fact or an idea in my research for an article that turns out to be useful in photography or in a radio programme and so on. And I never run out of things to do; there is always some interesting work waiting to be done.

WRITING THINGS DOWN

The world's first writers were the Sumerians, who in around 3000 BC lived in what is now Iraq. In order to keep records of trading and financial transactions, they invented a system of making wedge-shaped marks on clay tablets, many of which were baked hard either by the sun or in fires and have survived until this day. This script, now called cuneiform, has been deciphered by scholars and so we know a good deal about their civilisation. One of the rulers, King Hammurabi, had laws set in stone, literally, and displayed in public places; thanks to him we have the rather savage principle of an eye for an eye, first written down in about 1760 BC. There also survives the epic of Gilgamesh, a tremendous tale of the hero's search for immortality.

Cuneiform script – the first writing.

Later, when papyrus and paper became available, people began writing with pen and ink. The pens were originally made from quills – big feathers, often from goose wings, which were sharpened to a point with a knife. They were rather soft and needed frequent sharpening, but went on being used until the 1830s, when Joseph Gillott, in Birmingham, worked out how to make an improved steel pen nib that could be attached to a wooden handle. Steel pens already existed, but they were expensive at more than 10p each. He perfected the design of the nib and its manufacture, so that it became not only cheaper to produce but also as easy to use as a quill, and much more durable. He originally charged 5p per nib, but was so successful that he eventually employed 450 people and reduced the price to a hundred nibs for 1p, which had the effect of democratising writing – making it available to everyone, rather than just the rich. For the next hundred years most of the world's pens were made in Birmingham.

A reed pen, as used until around the sixth century AD.

Meanwhile, Gillott amassed suitcases full of money and was puzzled about what to do with it. He was interested in the arts and liked the landscapes of the artist J. M. W. Turner, who was talented but poor, so he visited the man in his studio and offered to buy one of his paintings. Turner brusquely refused, whereupon Gillott waved his arm at all the paintings round the room and asked how much Turner wanted for the lot. Turner was nonplussed. He named a ludicrously high price, but Gillott simply agreed and gave him cash on the spot.

Steel pen nibs dominated writing until the 1950s: when I was at school, we had dip pens, and inkwells in our desks, and an ink monitor in each class, whose job each morning was to fill all the inkwells from a big bottle of ink. Then there appeared the first fountain pens, with a bladder that contained enough ink for several hours of writing, and ball-point pens, invented by the Hungarian journalist Laszlo Biro, whose name has passed into the language.

The entrance to Grand Pipe, home of the best graphite deposit in the world.

PENCILS

During the Middle Ages, artists and draftsmen – including Leonardo Da Vinci – liked to make sketches with a dry marker before committing themselves in ink. They often used silverpoint or leadpoint – scribers with metallic points that left a faint line when drawn across the paper. They also used charcoal, but this gave a solid black line that was not easy to rub out.

In about 1560 a revolution in writing materials began with the discovery of a strange shiny black material in a hill in the Lake District. The story goes that shepherds sheltering from a terrible storm took refuge behind a fallen tree, and in the morning light found this mysterious black solid exposed among the roots. It left black marks on everything it touched, and it proved to be particularly good for marking sheep, since the black marks did not wash off in the rain.

The black material was shiny and hard, and yet somehow greasy on the surface. It came to be called black lead, plumbago, or, more commonly, wad; today we call it graphite, from the Greek word 'graphein', meaning 'to write'. Graphite

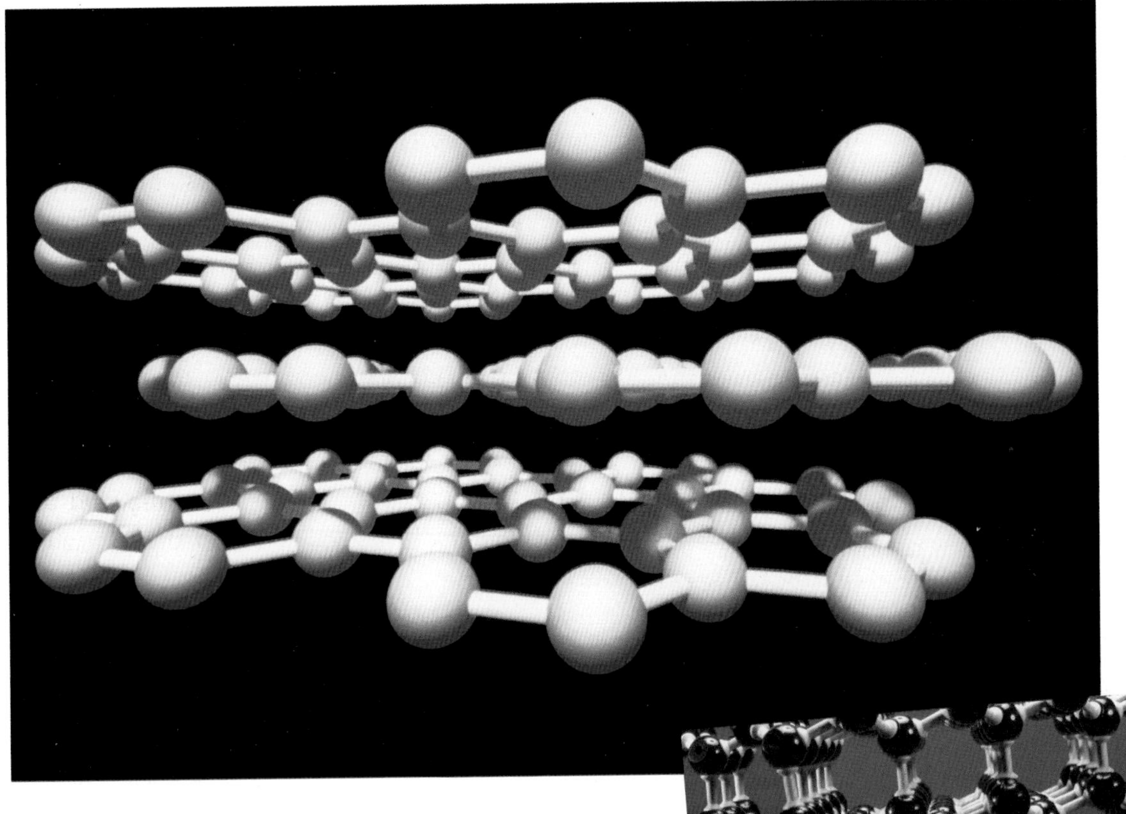

is actually nothing to do with lead, but a form of pure carbon. Diamond is also a form of carbon in which the atoms are bonded together in a vast three-dimensional structure that makes it immensely strong and hard. In graphite the atoms are arranged in flat sheets, rather like the shape of chicken wire, and because the sheets can slide over one another, the graphite feels greasy and can slide on to the paper to make marks.

The atomic structure of diamond (above) is a three-dimensional network, while graphite (top) has the carbon atoms in flat sheets.

The locals sawed lumps of wad into sticks, wrapped them in string or sheepskin to keep the marks off their hands, and sold them to others. Soon there was a considerable demand, and they began to dig into the hill to extract more. Demand jumped again when the wad turned out to be ideal material for lining the moulds for making cannonballs, since it could be carved into the right shape – two hollow hemispheres – but did not melt when the red-hot iron was poured in. The mine was known as Grand Pipe, and nowhere else in the world has there ever been found a

deposit of such pure graphite as that single hole in the side of Seathwaite Fell in Borrowdale. The mine was taken over by the Crown because of its importance for the making of cannonballs, and the royal agents used to flood the mine for years at a time to prevent anyone else getting hold of the stuff, but locals still looted the wad to make writing implements.

Eventually, someone – perhaps an Italian – devised a way of encasing a thin cylinder of wad in a thicker cylinder of wood and the 'lead pencil' was born. The name is doubly wrong. The 'black lead' in the middle is not lead at all but graphite, and meanwhile the first meaning of the word 'pencil' in the *Oxford English Dictionary* is 'an artist's paintbrush of camel's hair…'. The word comes from the Latin word *penicillum,* which to the Romans was a paintbrush and was in turn a diminutive version of *penis,* meaning a 'tail'. The second meaning is 'an instrument for marking, drawing or writing…'.

Most pencils today are hexagonal in cross-section, which has the advantage of being less wasteful of wood and also helps to prevent them from rolling off the table. The wooden case is made in two halves and then glued round the functional graphite rod in the middle.

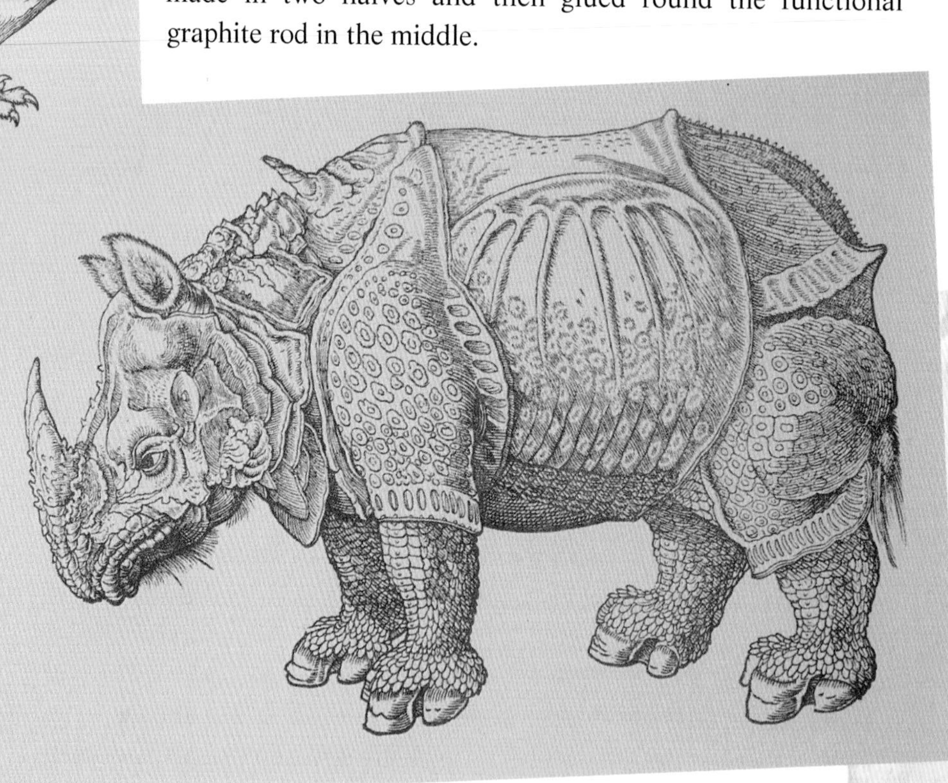

Drawings by Swiss naturalist Conrad Gessner, the first person to write about the pencil, in 1565.

Generally, the rod is made of powdered graphite compressed into a cylinder with a certain amount of clay, and fired, then dipped in oil or molten wax, which is absorbed into cracks in the solid and helps the pencil point slide smoothly over the paper. The clay makes the pencil point harder, but the mark fainter. The normal writing pencil is called HB, meaning 'hard and black'. Harder pencils are 2H, 3H, 4H and so on, while softer and blacker ones are 2B, 3B, 4B and so on up to 9B.

There is a story, probably apocryphal, that NASA spent $10 million developing a pen that would write in zero gravity, since all ordinary pens rely on gravity to keep the ink flowing. The Russians issued their cosmonauts with pencils.

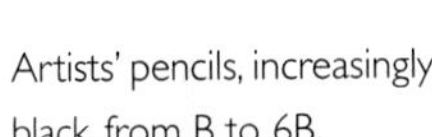

Artists' pencils, increasingly black, from B to 6B.

PAPER

The Sumerians wrote on clay tablets, and so did the Romans, who also used chalk on slate, and ink on thin sheets of wood. The Egyptians used papyrus, pale thin sheets made from reeds that grow on the banks of the Nile. Most people, however, switched when they could to paper, which is the most user-friendly and durable material that is relatively easy to make.

The word 'paper' comes from papyrus, but actual paper as we know it seems to have been invented by the Chinese, in around 100 BC, and they no doubt used it for recording their accounts and deals. The paper was made then, as it is today, by making pulp from water with softwood, or rags of cotton or hemp. This process separates individual fibres, which may then be bleached to make white paper. The suspension of fibres in water is poured on to a fine wire sieve, so that the water drains away and the fibres coagulate. As they dry, the fibres stick together and form a smooth writing surface.

Most paper is produced in a continuous wide roll, and because the water is flowing as the fibres settle, they tend to be lined up in one direction. This is why most paper is easy to tear in one direction (along the fibres) but much harder at right angles. Try it yourself; this is particularly obvious with newspaper.

Paper sizes vary from country to country, but in Britain the most common business sizes have a simple geometric relationship. Two sheets of A6 lying side by side just cover one sheet of A5; two sheets of A5 cover one sheet of A4 and so on. The largest size, A0, has an area of a square metre; so A1 has an area of 0.5 square metres, A2 0.25 square metres and so on. The downside of this neat relationship is that the actual sizes are not easy to remember – the common business size, A4, is 297 millimetres by 210 millimetres – and the ratio of length to breadth of each size is the square root of two, or 1.414.

The Ancient Chinese used paper not only as a writing material, but also for flags, kites and even armour. Even more important, they invented the idea of paper money. The story goes that traders along the silk road were continually robbed by brigands and so they suggested that instead of gold they should carry promissory notes – letters that promised to pay real money at a later date. These were worthless to the highwaymen and so the robberies were curtailed. These promissory notes were the ancient equivalent of cheques, but precisely the same technique was then taken up by treasurers. Today the British £20 note says, 'I promise to pay the bearer on demand the sum of twenty pounds' followed by the signature of the chief cashier of the Bank of England. I am not entirely convinced that the

The first promissory notes, or banknotes, must have looked something like this.

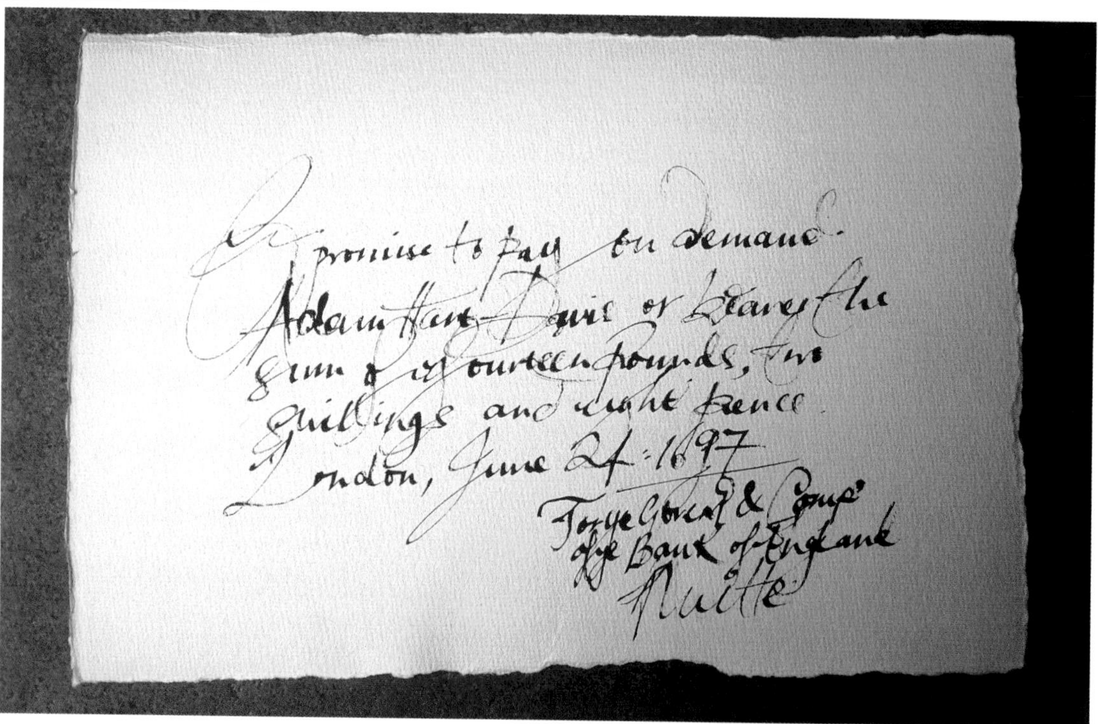

I promise to pay on demand
Adam Hart-Davis or bearer the
sum of [illegible] fourteen pounds, two
shillings and eight pence.
London, June 24 1697
For ye Govern & Comp
of ye Bank of England
[illegible]

Bank of England would actually give you the cash in gold on demand, but luckily everyone accepts these pieces of paper as real money.

This scheme first came to Britain in 1694, when William Paterson persuaded the King (William of Orange) to found the Bank of England. Paterson asked 1,500 rich people to lend the government money (at 8 per cent interest) and raised £1.2 million, which helped William to fund his war with France. Promissory notes were issued to those first lenders of cash, and these became the first banknotes in Britain. The loan was meant to last for only eighteen months, but in practice was never repaid, and is now known as the national debt. These banknotes had the effect of doubling the amount of money in the national coffers: for every £1 million worth of gold in the vaults, the bank could issue £1 million worth of paper promises, or banknotes – and then there would be £2 million of money. Today most people seem to operate on credit, so in effect the amount of money has trebled.

Those inventive Chinese were also the first people to make printing presses. In the West, books were laboriously copied out by hand, as manuscripts, but then in the second century AD the Chinese began to print prayers and other religious texts, using a carved wooden block for each page, which allowed them to mass-produce the texts. In the tenth century they tried to use movable type, with a single wooden block for each character, but this remained impractical for the Chinese language, with its thousands of different characters. However, printing with movable type became a major step forward in the West, when in 1440 Johannes Gutenberg began printing Bibles in Germany; a few of his Bibles have survived to this day, which is a tribute to the value of printing as a way of preserving ideas. William Caxton brought printing to England in 1476, with the translation of a French book called *Recuyell des Histoires de Troy*, because, he said, there was great demand for it: 'I have practysed and lerned at my grete charge & dispense to ordeyne this said book in prynte...that every man may have them attones' [at once]. The second book he produced was *The game and playe of the chesse*, which contained many fine woodcuts.

From the chess book printed by William Caxton in 1474, two years before he brought the printing press to England.

The fourth tractate & the last of the progression and draughtes of the forsayd playe of the chesse.

The first chapitre of the fourth tractate of the chesse borde in genere how it is made.

E haue deuiſed aboue the thinges that apperteyne vnto the formes of the cheſſe men and of theyr offices/ that is to wete as well of noble men as of the comyn peple/ than hit apperteyneth that we ſhold deuyſe ſhortly how they yſſue and goon oute of the places where they be ſette/ And firſt we ought to ſpeke of the forme and of the facion of the chequer after that hit repreſenteth and was made after/ For hyt was made after the forme of the cyte of Baby-

TYPEWRITERS

While I am at home, I spend much of the time struggling with office equipment, and I am constantly surprised at the way in which this has changed during my working life. When I got my first job in the publishing office of the Oxford University Press in 1971, I had nothing fancier than a pen and paper, although I did have a wonderful secretary, Meg Sheret, who came in every morning with her shorthand notebook and a pencil. I would dictate perhaps twenty letters and she would go off to the room she shared with five others and type the letters up on a large manual typewriter, making a copy at the same time using carbon paper.

The first commercial typewriter was patented by Christopher Latham Sholes in 1868, but Thomas Alva Edison and others soon muscled in.

This meant putting two sheets of paper into the typewriter with a flimsy sheet of carbon paper between them. Each time a character was hammered out by a key, the carbon paper would leave a fuzzy impression on the second sheet. Thus the secretaries made an office copy of every letter we wrote. Often they made two copies, one for the file, and one to send to someone else – an adviser or series editor, for example. All this

UNITED STATES PATENT OFFICE.

THOMAS A. EDISON, OF NEWARK, NEW JERSEY, ASSIGNOR TO HIMSELF AND GEORGE HARRINGTON, OF WASHINGTON, D. C.

IMPROVEMENT IN TYPE-WRITING MACHINES.

Specification forming part of Letters Patent No. **133,841**, dated December 10, 1872.

To all whom it may concern:

Be it known that I, THOMAS A. EDISON, of Newark, in the county of Essex and State of New Jersey, have invented and made an Improvement in Printing-Machines; and the following is declared to be a correct description of the same.

This invention is for printing by a type-wheel in a line upon a sheet or web of paper and then moving such paper along so as to print upon the line below. This invention is divided into the following principal features: First, mechanism for arresting a revolving type-wheel with the designated letter in position to be printed; second, the means for moving the type-wheel along between one impression and the next; third, mechanism for bringing the type-wheel back from the end of one line so as to commence at the beginning of the next; fourth, the devices for impressing the paper on the type-wheel; fifth, the feeding devices that move the paper the distance between one line and the next.

By moving the type-wheel along the line and across the paper the parts are simplified and rendered more compact than in those machines in which the paper has been moved; hence a roll or web of paper can be employed, and a telegraphic message printed thereon by hand, and cut off, instead of writing out the same, as now usual.

In the drawing, Figure 1 is a plan of the operative parts of the machine and part of the keys. Fig. 2 is a section at the line *x x*, near the type-wheel. Fig. 3 is a section at the line *y y*, representing the keys.

The type-wheel *a* is upon a sleeve that can be slipped endwise upon the shaft *b*. The guide-rods *c c* that are secured to heads upon the shaft *b*, and are paralled to such shaft, pass through holes in the type-wheel, and serve to rotate the type-wheel, but they do not interfere with the movement of the type-wheel and its sleeve endwise of the shaft by the means hereafter stated. The shaft *b* is of any desired length so as to pass over the range of finger-keys, and this range of finger-keys has the letters or characters corresponding to those on the type-wheel, and also the necessary keys for spaces between words and for moving the paper along from one line to the next. A pulley, continuously revolving by competent power, is applied to the shaft *b*, and an interposed friction allows the wheel to continue its revolution while the shaft and type-wheel are stopped. Upon the type-wheel shaft *b* are projecting pins or blocks 2, arranged spirally, or positioned so that when the stop-pin 3 is brought into the path of such block 2 the shaft *b* will be arrested by such pin 3, with the letter or character corresponding with the key depressed in position for printing.

The means for moving the pin 3 by the key might be varied; but I have shown the key *d* as acting upon a vertical bar, d^1, that has a pin acting in a cam-jaw, 4, upon the shaft *e* that carries such pin 3; hence, upon the depression of any one key the pin 3 connected with that key will be moved into the path of the block 2 upon the shaft *b*, and properly stop the type-wheel.

Springs d^2 are employed for raising the keys, and the key will rise slightly without liberating the block 2, in order that there may be time for the paper to be drawn away from the type, as hereafter described, before the type-wheel is again revolved. The finger-keys, for convenience, may be in two ranges, as shown. Beneath the range of finger-keys is a bar, *f*, supported by arms from the shaft f^1, so that when any one of the finger-keys is depressed the bar *f* will be moved, and, by the arm and pin 6, operate the feeding-bar *g*, which is made as a forked inclined pawl, 8, (see Figs. 2 and 4,) at the upper end, that is pressed between the spacing-pins 7 on the rack-bar *h*, and moves such rack-bar and the type-wheel along one space each time a key is depressed, and a spring, 9, allows this pawl 8 to yield as the bar *g* descends, and then springs back so as to take behind the next spacing-pin. (See Fig. 4.) The lever g' is placed above, as a finger-key, or connected with the bar *f* so that the pawl 8 can be operated to move the rack-bar and type-wheel between one word and the next without striking either of the letter-keys *d*. The rack-bar *h* is made to slide in the frames a' a', and is provided with a forked arm, h', that sets over the edges of a disk, 10, that is connected with the type-wheel *a*, so that said type-wheel *a* will be free to revolve, but will be moved along upon the shaft *b*, as aforesaid, by the rack-bar *h*. This rack-bar *h* carries, also, the inking-roller or wheel *k* that supplies the necessary ink to the type. In the surface of the rack-bar *h* are notches, and a spring-pawl,

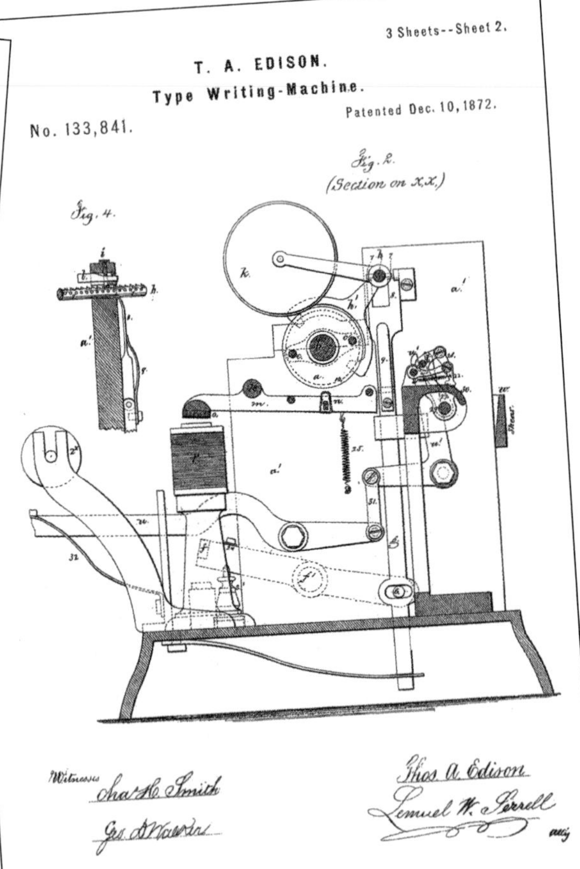

3 Sheets--Sheet 1.

T. A. EDISON.

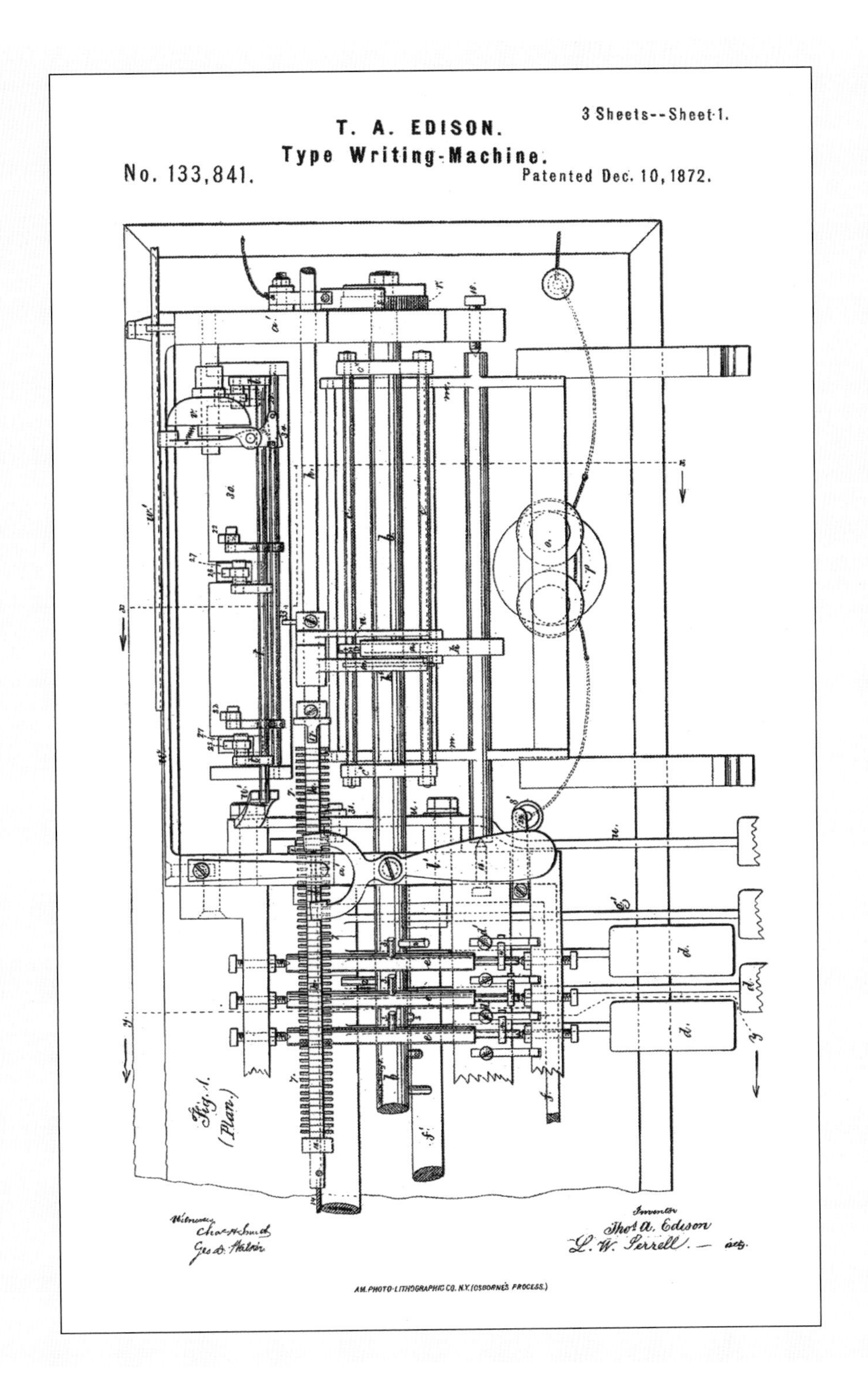

Type Writing-Machine.

No. 133,841. Patented Dec. 10, 1872.

Fig. 1. (Plan.)

Witnesses
Chas H Smith
Geo D. Walker

Inventor
Thos A. Edison
L. W. Serrell. — atty.

AM. PHOTO-LITHOGRAPHIC CO. N.Y. (OSBORNE'S PROCESS.)

had to be done with carbons, since the photocopier had not yet become a normal piece of office equipment.

Apart from the fuzziness, the real problem with carbon copies became all too obvious if you were not a brilliant typist. Every single mistake – if you typed 'teh' instead of 'the', for example – was faithfully copied by the carbon, so when you spotted such a mistake, you had to go carefully back, paint over the wrong letters on each sheet with white correcting fluid, and then retype. Horribly laborious and slow.

In the days of typewriters, copies of documents were made using carbon paper.

My own first typewriter was a tiny Remington portable that I bought at school. When I left the OUP in 1977, I bought from them a huge ancient Imperial that they were about to retire; it produced beautiful type but weighed about 10 kilograms and occupied most of my desk.

Computers did exist in those days, but only just, and certainly not for writing text. I remember the first computer that reached the University of York while I was doing my DPhil, in around 1967. We were allowed to use it for mathematical calculations, but we had to write our own programs, punch them all out on a stack of cards and hand the cards in over the counter. The next day we could go back in the hope of getting an answer – although usually I found I had made one error in the punching and the program would not run.

COMPUTERS

The computer is only as old as I am, but it has become such a universal tool that I now have difficulty imagining life without it. The first computers were built in the UK and the US in the early 1940s during the Second World War; the British machine was constructed primarily to crack the secret German codes, while the American machine was built to calculate trajectories of shells. The American company IBM saw the commercial potential, although in 1943 the company president, Thomas Watson, is alleged to have said, 'I think there is a world market for about five computers' – but this may in fact have been wrongly attributed to him.

As a lad, George Parker Bidder was a superb 'computer'; he went on to become a fine engineer, built the first railway swing bridge, and helped to complete the Clifton Suspension Bridge after Brunel's death in 1859.

The first adding machine was made in France in the early 1640s by mathematician Blaise Pascal, and an ambitious 'difference engine' was designed by the irascible Charles Babbage in the 1820s. As a student at Cambridge, Babbage had been annoyed by the errors that always seemed

By permission of the Rev. the Vice-Chancellor and the Worshipful the Mayor.

THE Nobility and Gentry of the University and City of Oxford and its Vicinity, are most respectfully informed, that the wonderful Phenomenon

Master G. Bidder,

A NATIVE OF DEVONSHIRE,

ONLY TEN YEARS OF AGE,

Will exhibit his amazing Powers of MENTAL CALCULATION, without the aid of Figures (of which he has no knowledge), at

A ROOM AT THE CROSS INN,

IN THE CORN-MARKET,

On MONDAY, MARCH 10th, and a few following Days,

From Eleven o'clock till Four in the Afternoon.

He has had the honour of exhibiting before her Majesty, and the Dukes of Kent and Sussex, Earl Stanhope, Sir Joseph Banks, the Lord Mayor of London, and many other persons of the first distinction in the Kingdom.

Ladies and Gentlemen wishing to put any numerical Question, are respectfully requested to come prepared.

The following Question was put by the Queen:

How many days would a Snail be creeping, at the rate of 8 feet per day, from the Land's End, in Cornwall, to Farret's Head, in Scotland, the distance by admeasurement being 838 miles.

He answered—553,080.

To the Question of his Royal Highness the Duke of Kent:

Multiply 7,953 by 4,648. Answer, 36,965,544.

The following Questions, among a great variety of others of the same nature, have been correctly answered by him.

Suppose a city to be illuminated by 9,999 lamps, and each lamp to consume one pint of oil in every 4 hours, how many gallons will they consume in 40 years?—Answer, 109,489,050.

Suppose the National Debt to be one thousand millions of money, in one pound notes, how long would it take a man to reckon the same, at the rate of one hundred per minute without intermission?—Answer, 19 years, 4 days, 16 hours, 40 minutes.

Suppose an acre of ground to contain 256 trees, each tree with 53 limbs, each limb 106 twigs, each twig 45 leaves, how many leaves are there in all?—Answer, 119,669,760.—What is the cube root of 673,373,097,125?—Answer, 8,765.

The promptitude with which he answers the difficult numerical Questions proposed to him, is not less extraordinary than his correctness; and those who witness his performance will be completely satisfied that there cannot be any deception in the case, but that he succeeds by the natural and unassisted powers of his own mind alone.

Mr. WOOLFORD, who has the care of him, challenges any one to produce his equal, for 500 GUINEAS.

Admittance—Ladies and Gentlemen, 2s. 6d. each; Young Persons under 14 years of age, 1s. each.

Tickets of Admission to be had of Messrs. Munday and Slatter.

*** Master Bidder will exhibit his astonishing powers of Mental Calculation to private parties, on application to Mr. Woolford, at Messrs. Munday and Slatter's, High-street.

[Munday and Slatter, Printers, Oxford.

Charles Babbage's prototype difference engine.

to appear in mathematical tables. The tables were calculated by people (called 'computers') who did all the sums with pencil and paper. Babbage reckoned he could build a machine to do the calculations and so eliminate errors, because people make mistakes, but machines don't. Unfortunately, the money ran out, and Babbage got into terrible disputes with his engineer, Joseph Clement, so his vast array of brass cogs and spindles was never completed; we shall never know whether it would have worked. The craftsmen at the Science Museum in London have built a full-size difference engine to Babbage's specifications, and it just about works, although both times I have watched it in operation it jammed.

Babbage has been called the father of the computer, but for me, the real pioneer was George Boole. Born in Lincoln in 1815, Boole was the son of an incompetent shoemaker who went broke, so at the age of fifteen George had to get a job to support his family. He could not find one in Lincoln, so he walked up the Great North Road to Doncaster and found a position as assistant teacher in a school in South Parade. He hated Doncaster, wrote home to say that no one there made gooseberry pies as good as his mum's, and spent his spare time studying mathematics; he had little money and therefore deliberately bought the books that he found most difficult to read, which were maths textbooks.

One frosty morning in January 1833 he went for a walk on Town Fields, opposite the school, and had what he described as a revelation – like Saul on the road to Damascus. He reckoned that if the working of the world can be described by Newtonian algebra, then the working of the brain should be describable by some other algebra. He thought he had cracked the mystery of the human mind. In practice, that was not true, but he had invented a new kind of algebra, which is now called Boolean algebra in his honour. It's a way of adding up logical statements to get a correct answer. At the time he became famous as a logician, but a hundred years later, when the first computers were being built, the builders realised that Boolean algebra was exactly what they needed. Now every computer, every mobile phone, even every washing machine

George Boole, the real father of the computer.

runs on Boolean algebra, and it all started on a frosty morning with a teenage teacher having a vision, in a field in Doncaster.

The first time I used a personal computer for writing – a BBC Micro with a view chip – was in the early 1980s. Until then I had written everything either by hand or using a manual typewriter. Having a computer with a word processor, so that I could correct mistakes and edit the text before printing, made writing much easier for me, and for millions of others. Today I use one all the time, and I would find it hard to go back. I find I can type (rather roughly) at about the same speed as I can think of the ideas and the order of the words. I write perhaps a paragraph, or perhaps a page, and then go back and tidy it up, correcting the mispritns, improving the grammar and the word order, sometimes adding bits or throwing chunks out completely. When I am firmly stuck into a book or a long article, I can usually write a thousand words a day, which means that if I had nothing else to do, the average book would take a couple of months. If possible, I go back a week or more later and read it again, to try to improve the grammar and the readability.

Mum and me trying to help Dad complete *The Times* crossword, just after I returned from India in the summer of 1962.

Forty-four years ago, while I was teaching in India, I used to write home to my parents every week. I found that on the air-letter forms then available I could get 900 words if I wrote by hand, or 1,000 if I typed right up to the edges with my little Remington. In one of those letters, I asked my father, who was then a leading London publisher, how to write good English. His reply was short and crisp: 'Use short sentences, and don't start them with "It".' I have followed this advice ever since – look at the text of this book – and it has served me well. When I was an editor at the OUP, I imposed the same rules on my authors, and none complained. However, when many years later I told my dad about his valuable advice, he said crossly, 'I never said any such thing.' One of us must have misremembered, but I still abide by those rules.

There is an argument that the word processor has destroyed well-crafted writing. In the days of pen and ink, authors had to think out each paragraph in its entirety before making a mark on the paper, then think it through again and then write it down. Now they can just slap down any old rubbish, knowing they can improve it later. I disagree. I still try to think through at least the sentence before I begin to type, and the ability to edit as often as I want gives me the confidence to get those ideas down before they evaporate.

One of the few people to have lost out in this revolution is the art historian, who in the past often revelled in the manuscripts of great authors, and could follow the tortuous process of writing by seeing which passages had been written, crossed out, rewritten and so on. Now this chain of evidence is almost entirely lost. My dad once reversed this loss. In 1960 he was chairman of the London Library and had to raise money to meet an unexpected tax demand. He asked all his author friends whether they could donate something for a literary auction. T. S. Eliot could think of nothing suitable. 'How about the manuscript of your most famous poem, *The Waste Land*?' Eliot gloomily replied, 'There isn't a manuscript: I typed it.' Then my dad had a brainwave; 'Right, Tom, if I give you a smart pen and some decent paper, will you write it out and make a manuscript?' And Eliot did just that; he went to Casablanca for the weekend, copied the entire poem from the Penguin edition, and sent my dad a telegram saying, 'O, CHAIRMAN, MY CHAIRMAN, THE FEARFUL TASK IS DONE' (a parody of Walt Whitman's line 'Oh, Captain, my captain, our fearful trip is done'). The new manuscript was duly bought by an American university for £2,800 and champagne was drunk in celebration.

Apart from the ease of editing, there are several other benefits of using a word processor. The computer checks my spelling and corrects simple mistakes automatically. It will count the number of words – important when I am supposed to write, say, 1,200 of them for an article. The final printing can be in any typeface, in any size and immaculate. I can keep an electronic copy, either in my computer or on a memory device, so the original cannot be lost. Loss of a manuscript used to be a terrible blow for an author – it happened to Isaac Newton and to Jilly Cooper, among others – but now it should never happen, although I have occasionally lost some work when the computer crashed. Perhaps best of all, I often don't have to

A computer sound card: amazingly, the sound produced by a computer is generated by this group of chips.

make a hard copy. I sent the 'manuscripts' of my last two books to the publishers by email and never printed them out at all.

Email has become – for me at least – the most important form of telecommunication. On busy days I get so many email messages that I don't have time to reply to them all, but email is slightly more of a blessing than a curse. It is particularly useful for exchanging large chunks of text. When I want to send an article for publication, or a chapter of my next book for criticism, I do it by email. When producers want to send me scripts, they use email. This saves paper, and all the effort of finding envelopes and stamps, and going to the post office. As long as everything is working, an entire book can be transferred to my publisher in a minute or two, and with no errors. The text cannot be physically lost in the post, and if it does not reach the other end, I still have the original and can send it again. What is more, the text arrives in electronic form, so it can be edited directly by the receiver and returned for further comments or editing; nobody has to retype anything. The advantages over conventional mail are speed and convenience; the advantages over using the phone are that email is not instantly demanding – you can look for it when you are ready – it can carry a huge amount of information, and you can print a hard copy for perusal.

I use my computer for a few other things too. I make PowerPoint presentations for my various lectures – recently, I have spoken at London Zoo about dead animals ('Stuffed, Mummified and Pickled'), at the AGM of the British Toilet Association ('Pissing in the Wind'), at the Royal Institution in London ('Seventeenth-century Inventions'), to an audience of American Skeptics on a cruise ship off the west coast of Mexico ('Scientific Eye') and to a group of engineering students on 'How Engineers Change the World'. Each new talk needs a new presentation, and I enjoy putting them together – most of them just pictures with simple captions. Sometimes the organiser asks me for a transcript of my talk, but I never write down the words. I think about the talk while I am assembling the PowerPoint and then speak about the slides as they come up.

I use PhotoShop occasionally, and inexpertly, to adjust and improve my photographs, and since I have recently bought a digital camera, this is gradually becoming more important. I have never been interested in processing my own photographs, or in making my own prints, but I am becoming hooked on minor manipulation of my photographs – removing

unwanted reflections and taking out distracting things in the background, for example.

My computer is also a great research tool; I use it to access encyclopaedias, the *Oxford English Dictionary*, the Dictionary of National Biography online, and the Web. Using the World Wide Web for research is slightly risky, since a good deal of the information there is misleading or just plain wrong. There are some websites that I can trust – I don't think the American Dental Association is likely to lead me astray – but I never trust a website that I do not know. When I am not sure, I often go to several different sites to try to get a majority view. Even this can fail – for example, I wanted to find out when the first commercial mobile phones appeared, but the several sites I consulted gave a variety of different answers.

I keep my address book on my computer, and my diary, and I synchronise these frequently with my amazing palm computer, an iPAQ. This beast is literally smaller than the palm of my hand, and yet with a couple of books inside, plus all my contacts and diary, it still has 20 megabytes of spare memory, which is more than three hundred times the total memory of my first computer, the BBC Micro. I use my little iPAQ to write on the train, which is what I am doing now (see page 109). During the last two days I have spent about eight hours on trains, and I find the atmosphere conducive to writing. Yesterday I wrote an article for a newspaper, and today I am working on this book.

FAX MACHINES

I remember clearly when the facsimile or fax machine came into its own. I was working at Yorkshire Television, in around 1978, and if we wanted to send hard copy in a hurry, we had to send a telegram or a telex. Telegrams were delivered in a few hours both by phone (if possible) and by a boy on a bike or a man in a van. Telex was quicker but also trickier. When I wanted to send one, I had to go and chat up the telex operator, who sat in a poky office in the heart of the building. She typed my message at high speed into her machine, along with the telex number of the destination. Then if all went well, the message would come out on a long, narrow strip of paper at the other end. It was not elegant, and would take only text – no pictures – but it was quick and cheap.

Then came fax, and suddenly from our own office room we could send a message to any other office in the world, as long as they too had a fax machine. It was like a photocopier, except that the copies came out somewhere else. And you could send pictures, handwritten notes, corrected text – anything that you could put on a sheet of paper. Knowing how fax changed our lives in around 1980; I was amazed to discover that the fax machine was first patented by Alexander Bain in 1843, which was thirty years before the telephone had been invented.

Sandy Bain was a shepherd's son, born in the remote north of Scotland, and was a dreamer both at school and in the field, where during the summer he was meant to be looking after the sheep. However, he was fascinated by clocks, and after making model clocks out of heather, was apprenticed to a clockmaker. At the age of twenty, he walked 21 miles from Wick to Thurso in the January snow to hear a lecture on 'Light, Heat, and the Electric Fluid'. From then on he was hooked on electricity. In 1841 he patented an electric clock, and in 1843 a device for sending pictures along telegraph wires – a clever mechanical precursor of the modern fax machine. Sadly, no one wanted to send pictures by wire, and Bain eventually died poor, but I cannot help feeling sorry for this genius of a shepherd's son who was 140 years ahead of his time.

The opening page of Sandy Bain's long and tedious patent for the first fax machine.

A.D. 1843 . . . . N° 9745.

Electric Time-pieces and Telegraphs.

BAIN'S SPECIFICATION.

TO ALL TO WHOM THESE PRESENTS SHALL COME, I, Alexander Bain, of 320 Oxford Street, London, in the County of Middlesex, Mechanist, send greeting.

WHEREAS Her present most Excellent Majesty Queen Victoria, by Her Royal Letters Patent under the Great Seal of Great Britain, bearing date at Westminster, the Twenty-seventh day of May, in the sixth year of Her

Office workers in Japan were particularly delighted at the advent of the fax machine, for Japanese is written in a complex mixture of phonetic *hiragana* and Chinese pictograms, called *kanji*, so it was almost impossible to type. Fax made interchange of information in Japanese much simpler, although today's word processors allow typing of all Japanese characters.

As a clock and fax inventor, Bain would no doubt have been amused by the experience of a friend of mine, an engineer working at MIT in Cambridge, Massachusetts, and collaborating with fellow engineers in Tokyo. They would sometimes send him information by fax at, say, 9 a.m. Tokyo time, and if he was still in the office he would receive it the previous evening, since Tokyo is fourteen hours ahead of Cambridge. Then he could write back, 'Thank you for your fax of tomorrow.'

TELEPHONES

I became a researcher in the science department at Yorkshire Television in 1977 and soon learned to use the telephone as a weapon. I found that with a little persistence I could reach almost anyone in the world, and in those days people were sufficiently excited by television that they would normally speak to me. One of my first tasks in the office was to find out how to get a film crew to the South Pole – a good challenge, given only a phone. In a couple of hours I discovered that the next flight there would not happen for nine months.

Later I had a long telephone conversation with one John O. Outwater, a professor of physics in Vermont, about why skates slide so easily over ice. A popular theory is that the pressure of the skate melts the ice to make a lubricating layer of water. The layer of water is correct, but the pressure isn't anything like high enough. In fact, what melts the ice is the heat generated by friction as the skate moves. He finally convinced me by drawing my attention to cross-country skiing. Basically, he said, you put your weight on the left foot and push the right forwards. If pressure melted the snow, then your right foot would stay still and the left would slide backwards, and all cross-country skiers would be permanently in reverse.

The telephone was invented by Alexander Graham Bell in 1875. Bell was a Scot, born in Edinburgh, became a teacher of deaf people and

apparently had the idea of transmitting speech by wire while working at a deaf school in Elgin, in a site now occupied by a Comet electrical store. He developed the system while working in Boston, USA, and got his patent application in only hours before his nearest rival.

The original telephones had what was essentially a loudspeaker at each end. Speakers make poor microphones, and the instrument was much improved by the carbon-granule microphone, invented – rather improbably – by Henry Hunnings, who was curate at a little Yorkshire village called Bolton Percy. Hunnings earned £1,000 for his invention, but ran off with a parlourmaid and disappeared from view. Bell invested in a vast sheep farm, which he called Sheepville, and spent years investigating the mistaken notion that sheep with many nipples would have more lambs than those more mammarily challenged.

Meanwhile, the telephone was slow to catch on; there was little incentive to invest in one when no one else had done so, and there was

Alexander Graham Bell's patent for the telephone.

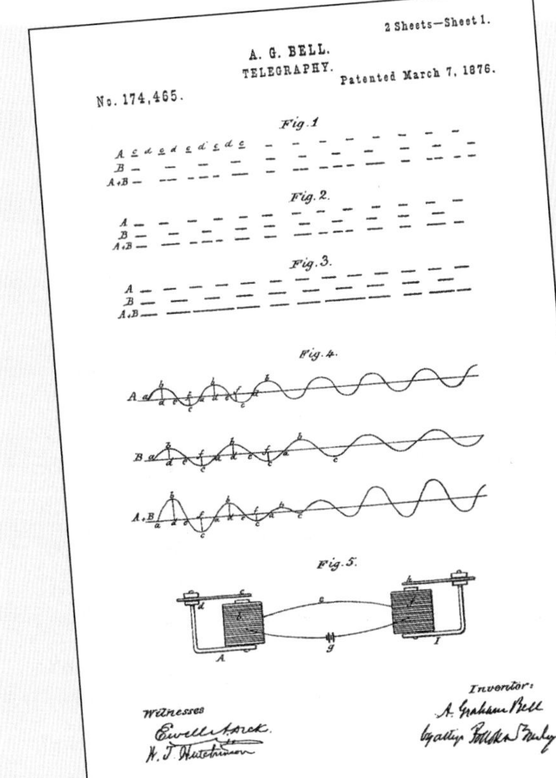

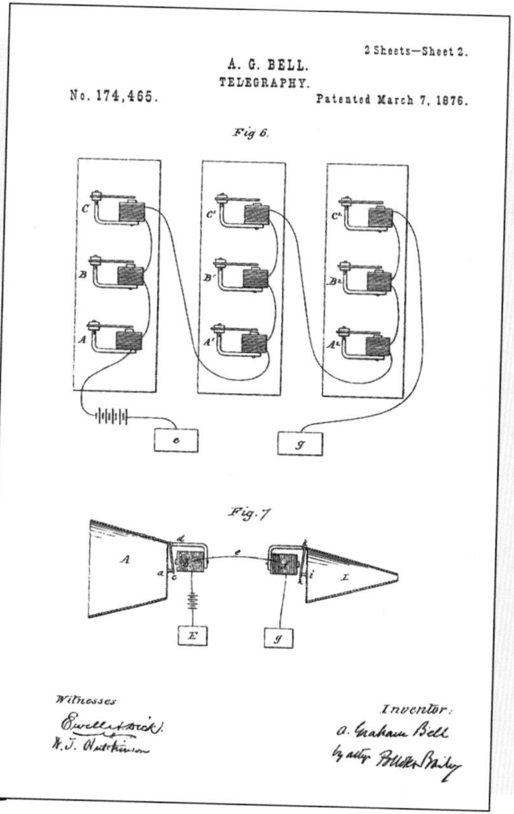

no one to talk to. But eventually it took off, and within a hundred years virtually every home and every office in the Western world had phones. Then in the early 1980s came the mobile phone – the first commercial one was the size and weight of a brick – but now most people in developed countries have their own personal phones. This is not unique to the developed countries, however; I was surprised on a recent trip to Kenya to find I had a good signal everywhere I went, and perhaps because landlines are scarce, mobile phones have become vital to large numbers of Africans. Indeed, this is probably how things will develop. Landlines are expensive to install and difficult to maintain, while mobile-phone towers are much easier and cheaper, and areas that have never had landlines will probably become served by mobile systems.

Today most phones are attached to people rather than buildings. Does this mean we are all being reduced to numbers?

Mobile phones are more sensible than landlines, since when you make a phone call, you usually want to reach a particular person, rather than a building, but what no one anticipated was that children would latch on to mobile phones and send one another text messages. Texting, or SMS, was added to mobile phones only as an afterthought, but is now the major use, especially for kids.

The mobile phone is indispensable for people working in radio or television. I remember struggling with public phones when I was out doing research, especially in the United States. The phone booths there would take no coins larger than a quarter (25¢), and a long-distance call might cost $5. Balancing a stack of twenty coins on top of the phone was almost impossible, and when I had several calls to make I was in trouble. Once, in Boise, Idaho, I checked into the Holiday Inn, made $100 worth of phone calls and checked out again, without even having a cup of coffee.

On another trip I reached New York on a Friday afternoon and had to find a couple of stories to film at the beginning of the following week, so I needed to work fast. I had already looked through the *New York Times* and found some leads, but I had to track down the people before they disappeared for the weekend. I went to my hotel in mid-town Manhattan, expecting to be able to go to my room and make my phone calls from there. However, the hotel computer had gone down and two hundred people were waiting to check in. So I went to the desk, acquired a bagful of quarters and did my research standing by the payphone at the back of the lobby.

Two great stories presented themselves. First, a man who trained athletes at the YMCA and could run a marathon in less than two and a

half hours, in spite of being fifty-eight years old and totally blind. Every morning he went running with his young blonde girlfriend, who used a dog lead to steer him round fire hydrants and parked cars.

The second story was of Curtis Brewer, who had been an athlete until a nasty illness had left him almost completely paralysed. He went to law school, getting other people to turn the pages of the textbooks, and became a lawyer and a champion of the rights of disabled people. I quickly tracked him down, and after chatting for five minutes on the phone, I had to ask him one question: 'Curtis, if you wanted to change the world, why didn't you go into politics?' His answer convinced me that we had a great story: 'Adam, I have a serious problem. I'm paralysed from the neck down. I cannot bend over to kiss ass.'

My favourite American phone booth is in Death Valley, up the northern end, near Ubehebe Crater. It's beside the Grapevine range of mountains, and the phone number is Grapevine 1.

Today, even though email has taken much of the strain of the acceleration of communication, the phone is still important. Yesterday I switched off my mobile before having lunch with an academic in a quiet university restaurant, but before the meal was over I had three messages, sent via various channels, from someone who wanted to speak to me urgently. These days I find it hard to escape completely from modern technology.

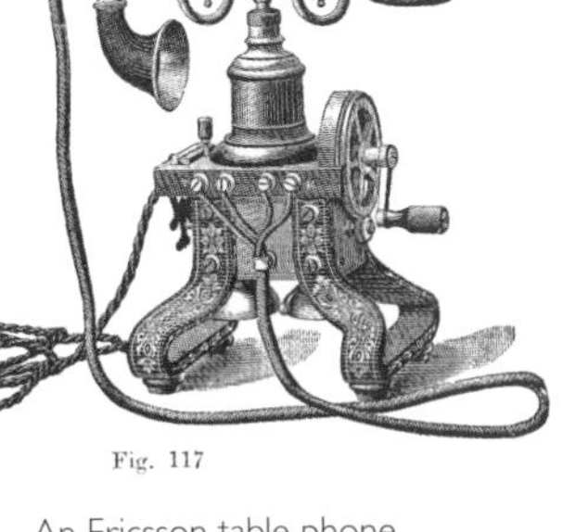

Fig. 117

An Ericsson table phone, 1892.

HOW PHONES WORK

When there were only two telephones in the world, there was no switching problem. Pick up one of them and you could get through only to the other. But as soon as there were three, there was always a choice to make, so the phones were given numbers, 1, 2 and 3. If your phone was number 1, then you could ask the operator for number 2 or number 3 and she would switch you through. Now there are millions of phones, and each has a unique number. Ordinary landline phones are connected to exchanges, each of which is connected to all the phones in that geographical area. These connections used to be by copper wires; today they are mostly by fibre-optic cables, which can carry far more information without having to amplify it so frequently.

In Britain, each exchange also has a number, so that someone dialling from a different exchange can ask for the right area. So if I want to phone the BBC at White City in London, I dial 0208 752 5252. Here, the first part, 0208, is the code for Greater London; 752 is the number of the White City exchange, and 5252 is the individual number of the BBC. So in principle each city can handle 1,000 exchanges (000 to 999), and each exchange can handle 10,000 numbers. Each city has a characteristic code: 0207 for Central London, 0121 for Birmingham, 0161 for Manchester and so on. In the United States, the system is similar: each number begins with an area code – 212 for New York, 213 for Los Angeles – although why these two codes are so close together when the cities are so far apart has always surprised me.

When you dial the number on your phone, it sends the number along the phone wires to your exchange and the message then gets passed to the correct city, to the correct exchange and to the correct phone, which then rings. The system works by simple switching. This used to be done by operators – I still remember that when I wanted to speak to my heart-throb Diana Brownrigg, I had to pick up the phone, wait for the operator and then ask for Checkendon 358. All the exchanges then had names. Today they have been replaced by numbers, and all the switching is done automatically.

Early telephone operators in Paris.

In a way, this is the story of the modern world: we have all been replaced by numbers. In particular, each of us has a mobile phone, and its number is our identity.

What seems like magic to me is how I can call a mobile phone. Suppose I want to call my son, who might be in London, Malta, or anywhere else on earth. How does the system know where to look? With landlines the system just follows the wires, but with mobiles there are no wires, and he could be anywhere. The answer is that it doesn't know where he is, unless he has told it. If he keeps his phone switched off, then it can't be found, but as soon as he switches it on, it sends out a radio signal, which is picked up by any phone mast within range, and that mast relays the signal by landline to a central computer. This process takes about ten seconds if you stay in the same country – try seeing how long your own phone takes to come up with the system name, such as UK Vodafone. However, it can take much longer if you switch it on in a different country. Once this connection is made, the computer then knows which 'cell' he is in. When I call, the computer passes the message directly to the mast nearest to my son, and from there it goes to his phone, which rings. In effect, each mast acts as a temporary exchange for its own cell, and if he is on the move, in a train, for example, the system will pass him on from one cell to the next as the signal strengths change. The masts seem to have a range of around a mile, although this depends on the lie of the land – less in towns and more in the country. There are anomalies, however; I was once filming on top of the BT Tower in London, 186 metres off the ground, and wanted to make a call, but found I had no signal. This surprised me, but the engineers there said you can never get a signal at the top. This must be a 'dead spot' in the network of masts; there will always be some dead spots in a complex array, where signals from two or more nearby masts are out of phase, and cancel one another out.

A mobile phone mast disguised as a tree.

Ordinary mobile phones exchange radio signals with those funny masts that keep springing up beside roads everywhere. For specialist use in remote places, and on ships, there are also satellite phones, which exchange signals with satellites in orbit around the earth. This means they need more powerful transmitters and better aerials than ordinary mobiles, and you have to carry a heavy box of stuff around with you. On the other hand, with a satellite phone you can make a call from the top of the BT Tower, and indeed from anywhere on earth.

Each mobile phone is a radio transmitter and receiver; they usually work at a frequency of 900 or 1800 megahertz, with aerials 8 centimetres or 4 centimetres long. The battery provides power of 1 or 2 watts, most of which is absorbed in your head. We do not know how the brain works, or whether this amount of power could damage it. There is no evidence for it yet, but advice in general is that young people, whose brains are still developing, should not use mobile phones too much.

PHOTOGRAPHY

When I have time at home, I fall back into photography, which often fills my thoughts; if I lie awake at night, I think about taking photographs. I am not interested in processing the film, or in marketing the results; my pleasure comes with the actual process of releasing the shutter. I particularly enjoy taking close-up pictures of slightly scientific subjects, including water drops, fibre optics and most of the photographs in this book. I take the majority of these pictures in my room – a former bedroom crammed with computer, books and camera gear.

My camera is usually mounted on a camera stand like a heavy tripod, which allows me to move it up, down and sideways by small amounts. Everything else is mounted on a Dexion frame, which is 2ft (60cm) square and 8ft (2.4m) high from floor to ceiling. The subject – say an array of toothbrushes (see page 37) – is placed on a suitable surface, or clamped to supports, on the frame. The lights – usually flashguns – are mounted on the frame, and one is connected to the camera, so that it fires when I release the shutter. The other flashguns have slave cells, so that they fire when the first one does. The background is often a piece of

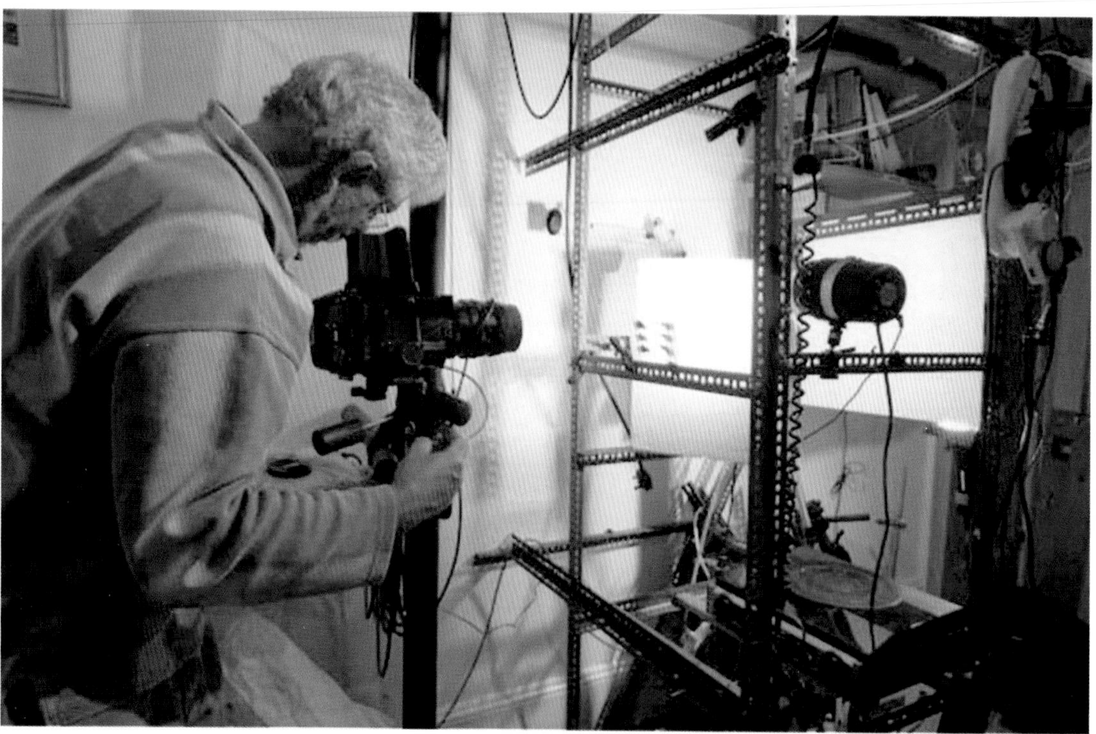

Using my Mamiya RB67 to shoot interdental brushes clipped to my Dexion frame.

translucent ('opal') Perspex, which I light separately from behind with more flashguns, so that I can make it any colour I want. For a black background, I use a piece of black velvet, which soaks up light like a sponge soaks up water.

The main reason for using the Dexion frame is that I can clip all these flashguns and the background to it, and I do not have to use an array of tripods and lighting stands, which would be all too easy to trip over. What's more, they will stay put while I take several pictures, perhaps with various background colours. The frame is held together by nuts and bolts, like a giant Meccano set, and I can add or remove bars in a minute, in order to provide supports for any kind of flashgun, reflector or subject. Taking close-up photographs, or macrophotography, demands patience and care. The depth of field is small and so focusing has to be precise, which is another good reason for having both camera and subject clamped in position.

The alarm clock in action.

Sometimes I try unusual lighting effects. I want to show the alarm clock in action, with the clapper vibrating and the clock itself teetering on one leg, as though dancing with the fury of the ring. So I balance it on one leg by using a long, stiff wire (a straightened coat hanger) poking through the background, set the clapper ringing and also leave the clock swaying on the wire. I light it from above with a powerful strobe lamp, firing fifty times per second, and give it an exposure of one-fifteenth of a second, in order to get multiple exposures and so show movement of the clapper.

Another tricky but rewarding thing to photograph is a drop of water bouncing on water. The water drips from an old chemical burette into a black plastic garden tray containing 2.5 centimetres of water, and I position the camera so that the edge of the photograph does not reach the edge of the tray; then my shot will have only water for background. Watching where the drops fall, I put a dummy object there and focus the camera. Some people think that in order to photograph water drops I must have a super-high-tech camera, but the opposite is the case. My workhorse camera is a Mamiya RB67, which I bought second-hand about twenty years ago. This beautifully simple machine is basically a box with a lens on the front and a film holder on the back. Everything is mechanical; it does not even have a battery.

A drop of water bouncing on a pool of water.

Water, like glass, is usually best lit from behind, so just out of shot I mount a vertical sheet of opal Perspex about 15 centimetres square.

Thirty centimetres behind that I mount a pair of flashguns, usually one above and one below, pointing at the back of the Perspex. I may also tape coloured gels over the front of the flashguns, or a gel or medley of gels over the back of the Perspex. These gels allow me to put whatever colour I wish into the picture.

Next, I mount a source of infrared radiation (like red light), so as to shine across the frame where the drop falls, and a triggering device on the other side. This is connected to the flashguns, so that as the drop falls it momentarily interrupts the infrared beam and so fires the flashguns. The trigger has a cunning delay mechanism, so that I can arrange for the flashguns to fire a few milliseconds after the interruption – just at the peak of the bounce.

When everything is organised, I black out the windows, turn off all the lights, open the camera shutter, switch on the flashguns and let one drop fall. Then I close the camera shutter.

This all takes some time, and I never know what I am getting, since I shoot these pictures on conventional film. So I first take pictures on instant-print film, to make sure I have the lighting right, the background clean and the timing as I want. I may shoot as many as ten Polaroid shots before changing to my favourite film, Fujichrome Provia, a positive film that produces colour transparencies or slides (whereas negative film produces negatives, from which prints can be made). And then I may shoot twenty pictures, varying the background colour, in that elusive chase for the perfect photograph. I must have shot hundreds of pictures of water drops, not to mention drops of milk, oil and syrup, and I am still looking for the perfect one. I have written more extensively about my photography in a previous book *Why Does a Ball Bounce?*.

Photography has a long history. Its roots lie in the discovery by Abu Ali Mohamed ibn al-Hasan ibn al-Haytham, otherwise known as al-Hazen, that light travels in straight lines. Al-Hazen was born in Basra in around AD 965 and went to Egypt to try to sort out the flooding of the Nile. The Caliph, al-Hakim, met him with open arms, and sent him up to Syene, now Aswan, to build a dam. Unfortunately, the Nile is still about half a mile wide there, and al-Hazen realised he had no hope with a few men and wheelbarrows, but he also knew the Caliph had a habit of executing anyone who annoyed him. So al-Hazen pretended to go mad, and stayed 'mad' for twelve years, until the Caliph died.

Al-Hazen had not wasted his time, however, but had secretly studied optics, and went on to write books on the subject. First, he explained how we see. The Ancient Greeks thought we see things by sending out rays from our eyes, but al-Hazen pointed out that the light is there anyway – you can feel it on your skin, and even on your eyelids when your eyes are shut. Open your eyes and the light pours in, like water into a plughole, and you see because of the light reflected from things all around you.

Al-Hazen also realised that we can't see round corners, which means that light travels in straight lines. This would also explain why shadows form behind solid objects. To prove that he was right, he built a small, dark room with a little hole in the door. He went inside and shut the door, and light streaming through the hole made an image on the opposite wall. This image was of the world outside, upside down and reversed left to right, but an accurate and moving picture of the world beyond the hole. People who saw this were utterly amazed: they had never before seen such an image. They thought it was magic, although actually al-Hazen was merely proving a scientific hunch. In particular, he showed that when there were three lamps outside, he got three images on the opposite wall and could extinguish any one of them by putting his hand between the image and the hole to cut off the rays of light. The small, dark room came to be called a 'camera obscura', which is Latin for 'small, dark room', and the name 'camera' now means any small box in which light comes through an opening and makes an image on the opposite side.

Tom Wedgwood, photography pioneer.

The camera obscura was for many years a piece of showman's equipment, used to amaze and delight audiences, but then artists realised how useful it could be to them, and there is some evidence that Jan Vermeer and a number of other artists in the seventeenth and eighteenth centuries used a camera obscura to help them get true perspective in their paintings.

Naturally, they wanted also to be able to capture automatically those images they could see in the camera obscura, and the first person to do that was Tom Wedgwood, the sickly son of potter Josiah Wedgwood. In around 1800 he discovered how to make photograms by painting white leather with silver-nitrate solution, leaving it to dry in the dark, then laying something like a leaf on top and exposing it to sunlight for some

Modern version of a Wedgwood photogram.

minutes, until the exposed leather had turned brown. Underneath the leaf was a lovely shadowgram of the leaf, where the sunlight had penetrated through. Unfortunately, Wedgwood never found a way of fixing these pictures, and they gradually darkened whenever they were exposed to light. Some were kept for a hundred years in books in the Royal Society, and people would go and look at them furtively by candlelight, in order to try to preserve them.

The next step forward was taken by John Herschel, son of the German musician and astronomer William Herschel. John discovered the cyanotype process, by which photograms could be made on ordinary paper previously painted with a solution of two chemicals. After exposure to sunlight, the uncovered paper turns royal blue, and when it is dark enough the leaf is removed and the paper washed to remove the remaining chemicals. The picture is now fixed and will last essentially for ever. Cyanotype pictures were used by Anna

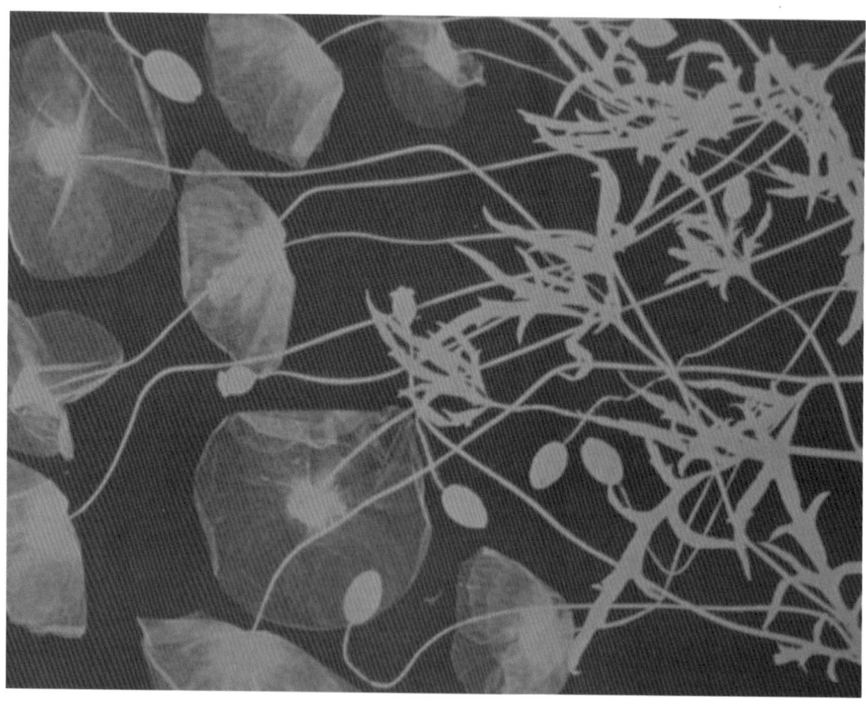

Anna Atkins with a copy of one of her photographs published in 1843.

One of Henry Talbot's original cameras.

Atkins to illustrate her book on seaweed in 1843; this was the first book ever to be illustrated with photographs.

Along came Henry Talbot, or William Henry Fox Talbot, as he liked to call himself, who took a couple more steps forward; he used a box with a lens in one side, like an advanced camera obscura, and discovered that using silver salts, like Tom Wedgwood, he could make a latent image with a much shorter exposure than his predecessors, and that although this image was negative – that is, sky and other bright parts of a scene came out black – he could develop this negative to the right density of greys and then use it to make positive prints. He described his system in *The Pencil of Nature* (1844), and thus was born the photographic process that lasted for 150 years, with some minor improvements including colour film.

Most of the film used today is colour film, which is sold in various speeds – ISO 50, 100, 200, 400, 1,000 and so on, where ISO stands for International Organization for Standardization. ISO 50 film is a slow film, which means that more light has to fall on it to make a properly exposed picture, but it will show fine detail. ISO 1,000 film needs little light, but will produce a grainy picture, in which fine detail may be lost.

The amount of light falling on the film depends on two things – the length of the exposure, which is often measured in small fractions of a

Oriel window, South Gallery, Lacock Abbey. This is my colour version of one of Talbot's earliest photographs, probably taken in 1835.

second – a 60th, a 125th or a 500th – and the size of the aperture through which the light enters the camera. The widest possible aperture is often f/2.8, which means the focal length of the lens divided by 2.8. To let in only half as much light, you need to close the aperture, or 'stop down' to f/4, and the next f-stops are 5.6, 8, 11, 16 and 22. Each of these successive stops corresponds to a halving of the area of the aperture. This means that an exposure of a 60th of a second at f/8 will let in roughly the same amount of light as an exposure of 250th of a second at f/4. Fully automatic cameras will choose sensible exposures for you, but using an ancient camera, or a modern one in manual mode, the photographer must make all the decisions. You might want to choose a short exposure to freeze movement, or a small aperture to give maximum depth of field – that is to get things in focus both close to the camera and further away. When you first acquire a camera, these things seem unnecessarily complicated and mysterious, but the more pictures you take, the better you get at taking not only photographs but also decisions about how to take them.

A few photographers still believe in black-and-white film – indeed, I love it for portraits – but most use colour. Negative film is the more common, and from colour negatives, prints can be made easily and quickly – that is what one-hour processing is all about. What is more, slight errors in exposure can often be rescued in the printing process. Colour positive film is processed to make transparencies, or slides, directly. There is no negative. That means that slight errors in exposure cannot be rectified; the result will always be too dark or too light. On the other hand, transparencies can be more brilliant and more true to the original colours than prints made from negatives, and that is why photo libraries prefer transparencies. In addition, they used to be used extensively for illustrated lectures, although this function has now generally been taken over by digital presentations.

Conventional film – positive, negative and black and white – is still in use, but it is gradually being superseded by digital photography, in which the light passing through the lens falls directly on to a photosensitive chip, and the intensity and colour of light falling on each pixel (or picture element) is recorded in digital form. Digital photographs still don't quite match the best photographs on film for sharpness and detail, and the best digital cameras are expensive, but the quality is going up and the price is coming down, and soon film will be a minor part of the industry.

FLOW

When I become seriously immersed in my photographic work – is it work, or is it a hobby? – I can lose track of time. If I am alone in the house, I forget meals; if someone else is cooking, I don't hear the dinner bell. Time just slides away. This is the nearest I come to what American psychologist Mihaly Cziksentmihaly calls 'flow'. He found that an artist sometimes becomes so immersed in her (or his) work that she is oblivious to the rest of the world; there is only herself and the painting. Indeed, it sometimes seems that she becomes one with the painting, and that all that exists is the process of slapping on the paint. This is not because she wants to finish the work and sell it, because often when it is finished she will just stack it against the studio wall and put another canvas on the easel. What she wants is to be utterly immersed in the painting process.

Portrait in oils by Ros Cuthbert: some artists become totally engrossed in the process of painting.

Cziksentmihaly is a friendly chap who explained some of this to me when he came to stay one night at our house. He said he had found the same effect with basketball players, with chess fanatics, and, perhaps more surprisingly, with climbers, who don't climb the mountain in order to reach the top; they go for the top in order to climb the mountain. What matters is the process of climbing. The

Robert Boyle, father of chemistry, born at Lismore Castle on the south coast of Ireland in 1627.

great pioneer of chemistry Robert Boyle must have felt much the same when he wrote, 'In my laboratory I find that water of Lethe [the river of forgetfulness] which causes that I forget everything but the joy of making experiments.'

Although I was a chemist for several years, I regret to say that I never felt quite so passionate about making experiments as Robert Boyle, but I do get passionate about taking photographs. Not only do I lose track of time, but I can become quite hot and sweaty, apparently through sheer concentration, since the process is hardly energetic. Out on location I have irritated my family and several radio and TV producers by insisting on taking just one more photograph of a building site, or a rusting lump of iron. I have been accused of never really seeing anything, because I am always peering through the viewfinder. I dispute this, and claim that not only do I see things clearly at the time, but I can see them later as well. I did once, while riding in a helicopter in Hawaii, bend down to reload my camera with film, and when I looked up I found we were within the crater of the live Kilauea volcano, with rock walls close all around, and red-hot molten lava below. I wondered for a second how we had got in there, and whether the rotor still had lift in the hot gases around us. But my camera does not often distract me from what is going on.

One difficult question is whether or not to lug a tripod around with me when I am filming, or making radio, or just walking. Using a tripod forces me to be more careful about my composition and prevents camera shake, so photographs taken with a tripod are generally better than those taken without; but it also slows me down, and tripods that are big and sturdy enough to be useful are heavy and awkward to carry. Generally, I do not take a tripod when we are filming or recording, since my first priority is to get the programme in the can, and slowing everything down with a tripod would not be fair to anyone.

Taking photographs on location in North America, 1980.

Why do I so love taking photographs? Partly because I enjoy teasing out the beauty in nature, and in human artefacts, but also partly because I am interested in all sorts of science and technology, and my photography allows me – indeed, forces me – to look at things more closely, and scientifically. Taking close-up pictures of icicles in the French Alps, I noticed that most had a column of tiny bubbles up the middle. I thought about this for some

weeks and realised why and how it must happen when I was photographing ice cubes in a drink. The photographs force me to look at things more carefully and more questioningly than I would otherwise.

In addition, my photography makes money: the Science Photo Library has more than a thousand of my pictures, which it sells for use in books, magazines and other places, and it sends me a modest cheque every quarter.

For most of my photographic career, I have used film, but I recently bought a good digital camera. Although the resolution is still not as good as old-fashioned film provides, the camera is a joy to use, and I suspect it may take over some of my life. As I learn to make best use of it, I realise that it does help me to get fine pictures. Indeed, there are a few in this book that I could not have achieved with film.

One difference between conventional film and digital is clear. Using conventional film, I take the pictures and then wait, sometimes for weeks, to get the transparencies back from the lab. I am usually disappointed with the results; the pictures are rarely as good as I had

The reflection in two mirrors (right) looks the right way round, but the single reflection (left) is reversed back to front (see pages 34–5).

hoped they would be. Then I have to decide whether to go back and shoot the pictures again, which is impossible if I was in a different country or if they were of some specific event. Using a digital camera, however, I can see instantly whether each picture is acceptable, and within minutes, if I am worried, I can transfer it to a laptop and look at it in detail. What is more, I can easily take dozens or even hundreds of photographs at the scene without having to spend lots of money on film, and without even having to waste time changing film, which on a medium-format camera takes a couple of minutes. I am more likely to get the critical shots, and the main problem is one of editing – throwing out the poorer shots so that they do not clog up all my computer memory.

In either case, I try to learn from my mistakes; indeed, that is almost the only way I have learned. Most of my photographs have mistakes, and I have taken hundreds of thousands; if I could really manage to learn something from each mistake, I should become a brilliant photographer.

COFFEE BREAK

When I am working at home, I take a coffee break at eleven, and this seems to be common practice in offices, at least in the UK. I find the break helpful to my work; I can concentrate for perhaps an hour at a time, and by eleven I have done two one-hour periods. The populist author Dan Brown says that he has an hourglass on his desk, and that after each hour he takes a break to do push-ups, sit-ups and some quick stretches, which help to keep his blood and his ideas flowing. I am less energetic, but when we are both at home, Sue and I play ping-pong five times a day, if possible: after breakfast, before coffee at eleven, after lunch, at teatime, and in the evening. Each game takes only about five minutes, but it gets the limbs moving, and – probably more important – raises the eyes from the computer screen. The 11 a.m. game is arguably the most useful, since it breaks up the morning's work, which tends to be the most intense. Our table is outside the kitchen door, so the game also gives us a breath of fresh air, and becomes interesting in gusty wind or in snow, but is much less fun in the rain.

In addition to the break from work, the coffee itself is a stimulant (see page 25) and so provides a bit of a kick to a flagging system. We drink only instant coffee at eleven, but some people insist on the real thing. In 1920 Samuel Prescott, professor of biology at MIT, set up the Coffee Research Laboratory there, with a $40,000 industrial grant. They fed rabbits vast quantities of coffee and found it did them no harm. They also made extensive investigations of the best way to brew coffee and concluded that the drip method was best, preferably in a glass or ceramic pot; the water should be just below boiling, and the coffee should be freshly ground.

RADIO

The first ever radio signal was sent some 55 metres from one building to another in Oxford, during the 1894 meeting of the British Association for the Advancement of Science. The sender of the Morse-code-like signal was Dr Alexander Muirhead, and the receiver was Oliver Lodge, a brilliant physicist who was knighted for his contribution to physics, but later diverted into spiritualism. Lodge was interested only in showing that radio signals could be transmitted; the idea was taken up with enthusiasm by the young Italian entrepreneur Guglielmo Marconi, who came to England and in 1896 transmitted signals several miles across Salisbury Plain. In 1901 he sent a radio signal across the Atlantic, and then the world really sat up and took notice. Only five years later the first radio broadcast was made, with voice and music coming from the loudspeaker.

A Marconi radio valve.

Radio waves are types of electromagnetic (EM) radiation, and form one end of the EM spectrum. At the most energetic end of the EM spectrum are gamma rays, followed by X-rays, then ultraviolet and visible light, infrared, microwaves and radio waves. Gamma rays have extremely short wavelengths – around a billionth of a centimetre, while radio waves have wavelengths from 10 centimetres to 2 kilometres; the BBC World Service broadcasts on 457 metres. EM radiation comes in packets of energy called photons, which travel through air, through empty space, and through us – there are radio waves going through your body at this very minute. To transmit a radio programme, the broadcaster sends out a strong carrier wave at a particular wavelength, modifying it by superimposing the programme on that basic wavelength. Your radio set then takes away the carrier, leaving the programme for you to hear.

Making radio shows is fun. I present about half a dozen programmes every year, usually half-hour documentaries about science or technology. I have worked with several excellent producers, and they do all the hard work – dreaming up the ideas, getting the research done, setting up the locations and the interviews, and telling me what to say. Sometimes we have a script – some presenters are uncomfortable without one – but more often we don't, and I am happy either way. On one memorable winter's day Mary Ward-Lowery dragged me to the bank of the River Tyne at Wylam, west of Newcastle, held the microphone under my nose and said, 'Tell us about George Stephenson's early life, and how he got interested in steam engines.'

For another programme, she made me climb hundreds of icy steps to the top of Lincoln Cathedral and describe the surrounding countryside, even though I hate heights, and could see nothing through the thick fog.

Speaking on the *Today* programme on BBC Radio 4 from a radio car outside my house.

Meanwhile, John Byrne took me to a clearing beside the River Wandle, in south-west London, to introduce a programme about water wheels, and as I began to speak, a robin flew up and perched on a twig within reach of my right hand. I am glad to say that it got into the programme, even though it was never in any sort of script. The joy of this type of documentary radio lies in its simplicity. You can do it anywhere. You need only a producer and a recording machine, and a good producer can tell at once whether what I have said will work in the programme. Many producers are prepared to suffer for the sake of their programmes; while making *Engineering Solutions,* Sarah Taylor not only took me to various scary building sites, tunnels and sewers, but was horribly seasick on our way out to see an off-shore wind farm.

The producer holds the microphone just in front of my mouth, and I speak in a slightly raised voice, as though I were addressing a classroom full of people. I slightly exaggerate any feelings I have of fear, excitement or amazement, to allow the listener to share the experience. At the same time I try to paint a word picture of what I can see: saying, 'Gosh, that's amazing!' isn't much use, but if I drop my voice to a loud whisper and say, 'Wow. A robin has just perched on a branch within reach of my right hand…' then I hope the listener can imagine the scene.

Radio cars are convenient but cramped.

Wind can be a problem, not because it is really noisy – it isn't usually – but because when it blows directly on to the microphone, it makes a distracting roar. The remedies are for me to face the wind so that the microphone faces away from it, and to put a windshield, like a furry squirrel, over the microphone. This solves the problem, because the wind no longer blows directly on to the microphone, but is stopped by the fur, which does not make a noise when it is ruffled. My voice still gets through because the fur does not provide a continuous barrier. When I am wearing a radio microphone for television, the sound recordist sometimes puts on a windshield and also tucks the whole microphone inside my coat, or top layer of clothes.

Another common problem is background noise. When you are recording on a building site, the occasional bang or clatter does not matter; indeed, it may add to the sense of being there. On the other hand, in a quiet country lane the sound of a helicopter, or someone strimming his edges, can utterly ruin the recording. Usually, the only solution is to wait until the noise stops.

Producer Sarah Taylor recording a radio documentary about the building of the Airbus 380 at Toulouse.

Is it useful or necessary to go out on location to make a radio programme, even though the listener will probably never know? Most documentary producers seem to think so. Even though all the background sounds from rain to buildings sites to blackbird song can be added in the studio, the sound of my voice varies subtly as we go from the open air into a shed or a room, and just being there gives me additional things to say. I would not have known that the Tyne was treacly-brown, that the cathedral steps were icy, that the robin had landed, or that Sarah would be seasick, unless I had actually been there, and perhaps that authenticity gets through to the listener.

Working in radio studios can be less fun, partly because they tend to be gloomy, grotty rooms without any sort of decoration. After all, they are designed to be soundproof rather than elegant. You sit on an office chair in front of a desk with a microphone either on a stand on the desk or hanging like an Anglepoise lamp in front of your nose. In a programme like *Loose Ends*, which has several contributors and is recorded as live – that is, in one continuous piece – we all sit round a long table, and each of us has a microphone. There is often also a band with multiple microphones on the other side of the studio. Working in the studio is

efficient, and you can get a lot done in a short time. I have taken part in a number of live programmes on various channels, and these are always entertaining. Being live on air adds a certain excitement; you have to be ready to respond instantly to anything that gets thrown at you, which is why politicians both love and fear such programmes as *Today*. Even documentary programmes recorded mainly on location generally have scripted sections, recorded in the studio, to tie together the location sections, to introduce some of the interviewees, and often to open and close the programme.

Speaking on BBC Radio Manchester from a Five Live studio in London.

TELEVISION

Television is more complicated. Filming (or technically recording) on location usually needs a team – camera operator, sound recordist, producer and/or director, and often a researcher. This means that you cannot sneak into anywhere you want and turn over (i.e. start recording) unnoticed. In a public place – the street or on wasteland – you usually get away with it, but on a train or in a house or a pub, you always need to get permission in advance, and then must try to avoid causing trouble by obstructing the way or disturbing the peace.

Technically, location recording is almost always done with only one camera, which is fairly lightweight. The cameraman – most camera operators are male – may mount it on a tripod, which gives steady, smooth shots, or may carry it on his shoulder, which may yield slightly jerky shots but provides much greater flexibility, since he can lean in to catch some detail in close-up, or actually walk with me into a building or from one room to another, or circle round me so that we can see the entire scene while I continue to speak to camera. Some cameramen also look after the sound, but more often a separate sound recordist puts a radio microphone in my clothes under my chin and then records sound both from that and from a separate microphone on a boom or pole, which is held just out of shot. The sound is recorded directly on to the videotape, which has the advantage that the words and pictures are locked together, and should always stay synchronised. The disadvantage is that the cameraman and the sound recordist have to be tied together by a piece of wire, which limits their mobility.

Because it is complicated, shooting documentary television is a slow process. Usually, we work all day, perhaps ten hours but sometimes sixteen, in order to get five or six minutes of good footage on tape. This means that a normal half-hour documentary takes about a week to shoot, but it can take even longer if there is a lot of travelling involved, or tricky things to capture. Later, it takes several days to edit, putting together all the best bits to make a continuous programme.

The work can be intense, and after rubbing shoulders for a week or more, the members of the team get to know each other quite well. The atmosphere can get tense when things go wrong – as they often do – and what is probably most important is that no one loses their temper, however frustrating the situation. People think television filming is

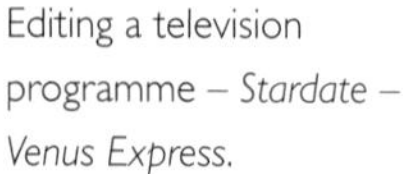

Editing a television programme – *Stardate – Venus Express*.

glamorous, but in fact you spend 99 per cent of the time travelling to the location or sitting around waiting – waiting for the interviewee to arrive, for the rain to stop, for the bus to move out of shot, for the man next door to stop mowing his orchard, and so on. On the other hand, television has taken me to wonderful places – to Land's End and John o'Groats, up Ben Nevis and Helvellyn, down the sewers in London, and to a dozen other countries, including Sri Lanka and Japan, which otherwise I might never have seen.

The best place I have visited was Death Valley, on the eastern edge of California. With Scottish director Charlie Flynn I drove there in a huge recreational vehicle, or RV, which we rented in Las Vegas. We bought a barbecue and plenty of chunks of meat, and set off across the desert. Our mission was to investigate the mystery of the moving stones of Racetrack Playa, a dried-up lake bed in the north-west corner of Death Valley. We camped there under the stars, Charlie cooked dinner, and I took photographs: there was a full moon, and Mars was in conjunction with Jupiter; in other words, they happened to appear close together in the sky, even though they were actually millions of miles apart in space. No doubt astrologers were excited by this coincidence. We must have been 30 miles from another human being, and the evening was a dream. When the sun came up in the morning, we were out on the hard sun-baked mud of the playa, looking for rocks that had moved. Sure enough, many seemed to have slid across the mud in nearly straight lines, leaving up to 100 metres of trail scratched in the mud behind them.

Steady as a rock? Racetrack Playa, Death Valley, California.

A rather less pleasurable experience was filming a piece about the history of nude bathing, at Scarborough on an icy-cold day in March. In front of the camera, I took off all my clothes (well, almost all of them) and ran down to the water, which turned out to be about 200 metres away, since the tide was rapidly going out.

The doctor who advocated nude bathing had said that total immersion was particularly effective in curing all sorts of ills, provided the water was cold enough. Well, the water was cold enough, I assure you. I had to splash out another 50 metres to where it was knee-deep. Then I lay down among the knobbly rocks to get right under the surface. The pain was severe. I sprang up again, spluttering, dripping and shivering, and said to the camera that the water was incredibly cold and therefore that I must be incredibly healthy. At this point, Charlie, who was cameraman on that occasion, said that I had been brilliant, but he was sorry, he had been laughing so much he had shaken the camera, and please would I do it again?

I have also, on behalf of television programmes, visited Sydney Opera House, the active crater of the volcano Kilauea on the Big Island of Hawaii, and the Pyramids of Egypt. The best experience I have ever had for television was dropping tomatoes off the Leaning Tower of Pisa – not many people get paid to do that. The Ancient Greeks alleged that big things fall faster than little ones, and that for example a brick will fall twice as fast as a half-brick. The great scientist Galileo Galilei, one of the first to do experiments, did not believe this. He was professor of mathematics in Pisa at the time, and so, in 1590, according to legend, he went up the Leaning Tower and dropped from the top light and heavy balls, in order to show that they fall at the same speed. I therefore followed in his footsteps by dropping tomatoes, a little cherry tomato with my right hand and a big beef tomato with my left.

Following in the footsteps of Galileo, in Pisa.

Dropping the two at precisely the same instant was quite difficult. The big one needed a tighter grip, and I found there was a danger that I would release it more slowly. However, I managed to do it right eventually, and the crew, on the ground, managed to film both tomatoes hitting the ground together. Intriguingly, I dropped them stalk-side uppermost, and they fell all the way down without tumbling and hit the ground still stalk-side up. We thought they would explode on impact, but in practice, they split neatly down the sides in several places, like Chinese lanterns, and made no mess at all.

Although this was great fun, and a satisfactory demonstration that big things do not necessarily fall faster than little ones, I was terrified at the top. There is a long, tedious climb up a spiral staircase, and then you emerge on the highest gallery, just below the campanile. The walkway is perhaps a metre wide, of polished marble, and sloping with the lean of the tower, so you could easily slip. There are awkward steps up to the campanile, which make a series of trip hazards. The handrail, rather rusty, has a gap under-

neath that is quite wide enough to slip through – and it is 50 metres straight down. I calculated that if I fell, I should – just like the tomatoes – take slightly more than three seconds to reach the ground – plenty of time for my life to flash before me...Watching the film, I can see my knuckles were white as I clung on to the handrail.

In armour for *What the Romans Did For Us.*

A few embarrassing things have happened to me while filming. In a Cornish port, the Royal National Lifeboat Institution rigged up a rope across the harbour to show me how a breeches buoy works; it's a sling someone can sit in to be hauled to safety from a sinking ship. The tide was coming in, and I asked whether I was going to get wet; the day was cold, and I was wearing my warmest clothes. If I was likely to get wet, I could have put on a wetsuit, which would have been sensible. 'No, no, you won't get wet – well, maybe your ankles.' I clambered into the breeches buoy, they hauled me slowly across, and the icy water came up to my armpits. As I said on TV afterwards, 'I know what got wet, and I don't call them ankles.'

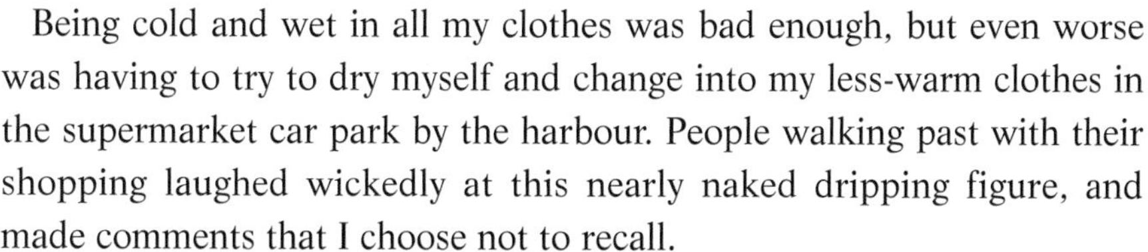

Being cold and wet in all my clothes was bad enough, but even worse was having to try to dry myself and change into my less-warm clothes in the supermarket car park by the harbour. People walking past with their shopping laughed wickedly at this nearly naked dripping figure, and made comments that I choose not to recall.

These events are fun, but rare. Most of the time filming on location is tedious, wet and cold – or occasionally, uncomfortably hot. I recently spent a day filming in Kenya's Rift Valley. The temperature was 38°C in the shade, but there was no shade. I have not worked in that sort of temperature for many years, and I had forgotten how it knocks the stuffing out of you; I must have drunk 8 litres of liquid during the day, which seemed a lot at the time, but apparently, I should have drunk twice as much. Severe dehydration is dangerous; if you lose 15 per cent of the water in your body, you are in danger of dying. The problem is that the working of your body depends on hundreds of chemical reactions going on all the time, and if you lose much water, the concentrations of all the chemicals changes enough to prevent those reactions from happening – so your body will just stop working.

Filming for Practical Action with Maasai people near Magadi in the Rift Valley, Kenya.

Recording television programmes in a studio is entirely different. For one thing, the place is air-conditioned, so although the lights can be hot, the temperature is generally pleasant. The atmosphere, on the other hand, can be tense. On location, you take the camera to the story; in the studio, the producers bring the stories to the cameras. Usually, you try to make an entire programme in one session, so a half-hour programme takes half an hour, rather than a week. This means that everything has to be well organised, so that it can happen in real time. Interviews are shot with two or more cameras, so that there are single shots of all those involved. For demonstrations, there will be one camera permanently on close-ups, so that viewers can see the details.

On the smooth concrete floor, the three or four heavy cameras are wheeled about on large wheeled pedestals (the cameras are called 'peds') by their operators – women quite often do this job in the studio. Researchers and technicians dart about, trying to keep out of the way. The stage manager makes sure everything is in the right place. The floor manager is in charge on the floor and tells people when to start or stop performing, while all these people have earphones, and are under the control of the director in the control room or gallery, which is often one or two floors higher in the building and out of sight of the studio. Communication is vital, and everyone has to follow strict discipline about not talking too much. Because those in the gallery cannot see the studio floor, they can find any delays highly frustrating. They don't realise that someone in the audience is ill, or a dog has appeared in the studio, or a piece of scenery has fallen over.

Filming for *Stardate* at the European Space Agency, Darmstadt, Germany.

Once the show begins, the director tells the camera operators which shots are needed and asks the vision mixer to cut from camera to camera, so that the entire programme is edited as it happens. Indeed, when the programme is transmitted live, as *Tomorrow's World* used to be, and *BBC Breakfast* is every morning, the show has to be put together as it happens, and if it goes wrong, then you just have to smile and carry on. Live television is fun, partly because the

adrenaline runs high, and you know you have only one chance to get it right.

In the studio, if I am doing a number of pieces, as I did in *Tomorrow's World*, I am happy to use autocue, also known as a teleprompter. This is a cunning device that allows an operator to project the script on to a sheet of glass in front of the camera, so that the presenter can read it while looking straight at the lens, and therefore at you, the viewer. This is how news readers manage to do long, complicated pieces to camera without forgetting names or dates.

Out on location, I rarely have to do more than a paragraph in one shot. I can remember that much, so I never use autocue. In practice, I don't learn the script, unless there is a quotation that I have to recite exactly. Instead, I read the script, take in the story or the point I am supposed to make, and then tell it in my own words. If we do two or three takes, the words come out differently on each take, and usually they get better each time. I have been known to need ten takes for a single piece, but usually only when things beyond my control interfere, or when I am so tired I cannot get a sentence out clearly.

Location can be a lonely place, especially in cold rain, but the studio is always a hive of activity. There is a large technical crew looking after lights, cameras, sound recording, props, scenery, and so on. There are often several producers and researchers, and sometimes interviewees or performers to be shepherded and kept happy. In order to make sure that it's all right on the night, there are usually rehearsals so that everyone knows where they are meant to go and at what time. So a single half-hour show may have the scenery moved in overnight. Rehearsals may start with a walk-through, or stagger-through, at 10 a.m. and continue for most of the day, and then the final show may be recorded or transmitted live at, say, 7 p.m.

More filming at ESA.

Television work is always tiring, and I have never understood why. When I was a researcher, I was bewildered on location – I felt I never knew what was going on. When I became a producer, I thought I knew what was happening, but felt I had no control. As a presenter, I am happy to give up both organisation and control, so that I can concentrate on my job of speaking to camera. But still I find I become exhausted by a day's filming or recording in the studio. Perhaps this is because I am concentrating all day on what I have to say. At any rate, I love the work and finish most TV days tired but happy.

MAKE-UP

Before I am allowed into television studios, make-up artists usually cover my face with pale-brown powder, or foundation, which covers up the worst of the wrinkles and scaly red patches, and prevents shine reflected from the studio lights. This foundation contains a mixture of iron and titanium oxides. Iron oxide comes in three different forms – yellow, brown (which is rust), and black – while titanium dioxide is pure white; the combination provides most of the 'natural' foundation colours.

I have never used lipstick, but I must have swallowed a fair amount; it is made from beeswax and castor oil, or a similar combination, with a metal-oxide pigment. Sparkly lipstick has flakes of mica or similar pearlescent materials. Women who use lipstick often have many colours available, from pale pink to dark purple. Three-quarters of those used are basic 'bullet' lipsticks, but the problem with them is that the stuff comes off on tissues, spaghetti, coffee cups and men. Kiss-proof lipsticks started by having less oil and more wax, which made them harder to apply but less easy to rub off. Now, however, there are two-stage lipsticks: first, paint on the colour with a brush, let the solvent evaporate, then cover with a wax topcoat to give luscious lips. I am told that this is flexible, feels good, and will not come off. Just top up the topcoat from time to time and it will last through a long evening.

Lip gloss is a flowing polymer that may have some mica for glitter. Don't wear it with long flowing hair, which will stick to it – and avoid wearing it when you are getting your hair cut. Kissing with lip gloss is a bad idea, because it feels sticky. Both lipstick and lip gloss are strictly non-poisonous, since quite a lot is bound to be eaten.

There seem to be almost no cosmetics aimed at men. Men started with aftershave and have moved on to moisturiser, but that seems to be as far as most chaps want to go. Women, however, have always wanted to look

more attractive. Five thousand years ago the Ancient Egyptians made their eyes look bigger and emphasised the eyelashes by using kohl – ground-up stibnite (antimony sulphide), lead sulphide, or soot mixed with something sticky and greasy such as sheep fat. One enthusiastic user of kohl was the biblical temptress Jezebel, although the Bible has warnings about women who paint their eyes. Today some 75 per cent of Western women use mascara and foundation. Mascara is a mixture of a dark pigment and a sticky base of wax and oil to hold it in place; so things have not changed much. Putting mascara on the eyelashes with a soft brush or 'wand' is a tricky operation, and I am told that women not only tend to screw up their faces into weird contortions while they are doing it, but sometimes sneeze, which blasts the mascara everywhere.

Cleopatra is said to have bathed in asses' milk because she thought it would make her skin smooth and milky white. For hundreds of years white skin was desirable, because it showed that you did not have to work out in the fields for your living, but were a lady – or gentleman – of leisure. To whiten their skin, people used chalk, rice flour or ground-up white lead,

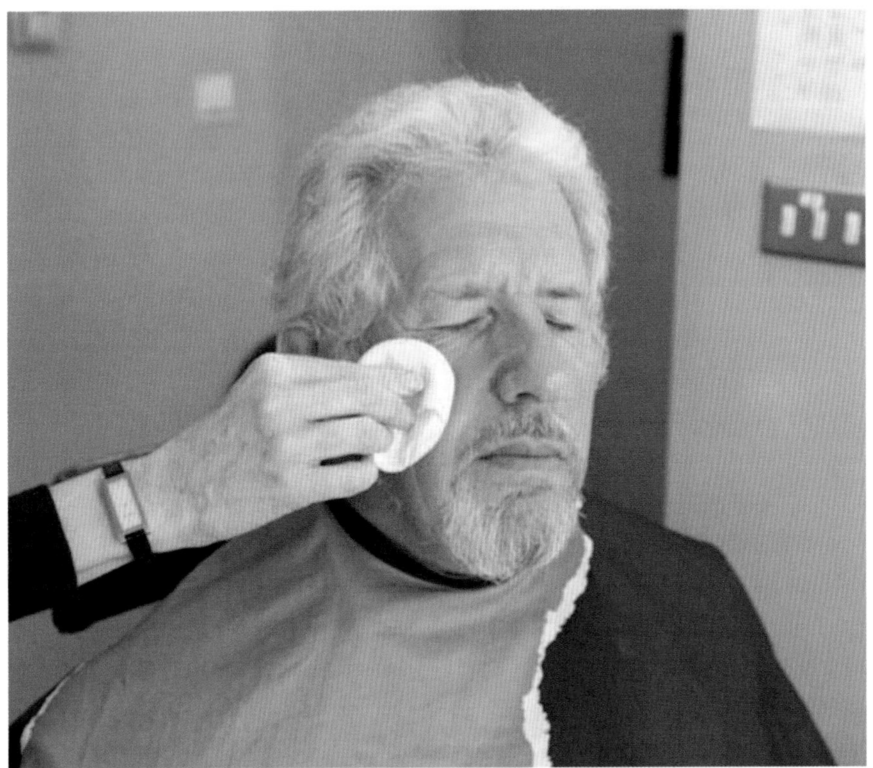

Being made up with foundation before going into the studio for *BBC Breakfast*.

not realising that lead is dangerously poisonous; it must have killed hundreds of users of make-up, and also many of those who kissed them.

During the Hollywood years of the 1930s, however, the ideal of having a milk-white skin began to give way to a new Western dream of a healthy suntan. Today pasty white skin is thought to be unhealthy; most people want to be gorgeously bronzed – despite the increasing incidence of skin cancer, as people spend too much time in the sun or on sunbeds. One solution to this problem is the fake tan. When I was a boy, we could turn our skin brown (for disguise) by rubbing it with the juice from the outer green case of walnuts. Nowadays you can buy fake tan in a bottle; the active ingredient is generally dihydroxyacetone, which becomes darker when it reacts with the amino acids in your skin. Go to a beauty salon and you can have it sprayed all over your body.

LUNCH

By the middle of the day I become a little hungry. Technically, what happens is that the level of glucose in my blood drops and I experience mild hypoglycaemia. Some people feel desperately hungry, but luckily I don't. This puzzles me, for I am fond of my food, generally eat too much, and have been overweight for most of my life; so I would expect to feel starving by lunchtime. But when I am at home immersed in photography or writing, I sometimes find the time has leaped on to 3 p.m. and I have missed lunch altogether. Occasionally while filming, the director tries to carry on without a lunch break and usually the crew protest vigorously. This happens most often in the cold of winter, when filming hours are short, because the light goes by 3.30 or 4 p.m. Although a midday break to warm up is immensely welcome, it is also costly in terms of filming time, and usually I am prepared to carry on without lunch if we are desperate to get a particular sequence recorded.

Many lifestyle magazines urge their readers to lose weight with the latest diet, but I am not keen on diets. I can well believe that if you eat nothing but grapefruit you will get thin, but what is to prevent you from putting all that weight on again when you get sick of grapefruit? The allegedly scientific diets may be championed by models and actresses – like those expensive skin creams – but they are usually heavily criticised

by nutrition experts. Whenever I am threatened with a diet I remember the story of William Banting, a chubby Victorian undertaker, who was only 165 centimetres (5 feet 5 inches tall), but by the time he was sixty-five weighed 92 kilograms (14 stone 6 pounds). He was distressed at his shape, at the frustration of being unable to tie his shoelaces, and at the fact that he 'puffed and blowed' when he walked upstairs. He even had to walk downstairs backwards.

Banting consulted several doctors and was given lots of advice. He swallowed gallons of patent medicines; he went rowing for hours every morning; he visited Turkish baths, and even tried the new-fangled practice of sea bathing, to no avail. Then, because he was going deaf, he went to Soho Square and consulted Mr William Harvey. Mr Harvey said his deafness was caused by corpulence, and that the remedy was to go on a diet. He told Banting not to eat bread, butter, milk, sugar, beer, soup, potatoes or beans, but to eat mainly lean meat, fish and dry toast. (Why toast was allowed, but not bread, is not clear to me.)

A slightly simplified version is printed on the next page. This is basically a high-protein, low-carbohydrate diet, rather like the modern Atkins diet. What astonishes me is the amount of food, and in particular the amount of alcohol, that Banting was expected to consume. Even I don't often drink a tumbler of brandy as a nightcap. And yet in spite of that enormous intake of alcohol, the diet was successful. Within a year, Banting lost more than 20 kilograms (3 stones), and he felt better than he had for twenty years. He was so delighted that in 1863 he wrote a pamphlet called *A Letter on Corpulence, Addressed to the Public.*

William Banting, the once corpulent undertaker.

This too was an immense success. Tens of thousands of copies were bought by others who wished to be slimmer. The word 'Banting' became synonymous with dieting, and the verb 'to bant' entered the English language – 'I say, you do look well! Are you banting, my dear?' As a result of this, William Banting became quite rich and enormously famous, and thousands of people followed his advice – which was rather unfair, really, since the advice came in the first place from Mr William Harvey of Soho Square.

This advice may have worked for the corpulent undertaker, but I do not believe it could work for me. I prefer to trust in simple thermodynamics: I need to burn more calories than I consume. In other words, I must exercise more and eat less.

BANTING DIET

Breakfast	4–6 oz (110–170 g) beef, mutton, kidneys, broiled fish, bacon, or any cold meat except pork 1 large cup of black tea 1–2 oz (30–60 g) dry toast
Dinner	10–12 oz (280–340 g) of any fish except salmon, any meat except pork 2 or 3 glasses of good claret, sherry or Madeira but NO champagne, port or beer any vegetable except potato 2 oz (60 g) dry toast fruit out of a pudding any kind of poultry or game
Tea	4–6 oz (110–170 g) fruit 1 large cup of black tea a rusk or two
Supper	6–8 oz (170–225 g) meat or fish, as dinner 1 or 2 glasses of claret
Nightcap	1 tumbler of gin, whisky or brandy

Why do I care about my weight? Partly because I am vain – I am told I look better when I am thinner – and partly because there are sound medical reasons: obese people are more at risk of injuries to ankles, knees and hips, and more likely to suffer heart attacks and strokes. In 1943 – the year I was born – the Metropolitan Life Insurance Company published a set of height-weight tables. The Met Life statisticians were interested in when people died, and these tables simply correlated longevity with weight. For men and for women, and for small, medium and large 'frame sizes', the tables gave 'desirable weights', which actually meant the weight at which people stayed alive longest. I am 183 centimetres (or 6 feet) tall and have a medium frame; so according to the tables, my desirable weight is between 75 and 80 kilograms (between 11 stone 8 pounds and 12 stone 6 pounds). I actually weigh more than 100

kilograms. I would like to get down to 90 kilograms, but losing weight is difficult.

The Metropolitan Life tables have largely gone out of fashion, and these days the 'in' words are Body Mass Index or BMI, but the result is the same; at over 100 kilograms I am obese, while if I can get below 100 kilograms, I shall merely be overweight.

When I was a publisher, I frequently took authors out to lunch, which was a good way of finding out how their writing was going and also providing a small perk – indeed, in one memorable case a teacher was persuaded to come and talk to me only because he had heard on the grapevine that I provided good lunches. I talked him into writing a book, and within three years he was a successful author and bought himself a Volvo from his royalties.

In those days I often had several good lunches a week – starter, main course, pudding, shared bottle of wine. Perhaps that is why I am now too fat. Good lunches rarely come my way today, which must be a good thing for my figure. Perhaps once a month I have a proper lunch with one of my agents, or with a television producer, but it is rare, and now that I am an author, publishers no longer seem to have time to give their authors lunch.

When I am at home, lunch is often salad – usually made by Sue, if she is at home too – lettuce, cucumber, tomato, spring onion, sweet pepper, with perhaps rocket or spinach from the garden, plus cashews or pine kernels or sunflower seeds, and a simple oil-and-vinegar dressing. This

A rainbow salad, with as much as possible fresh from the garden.

may be accompanied by a piece of bread and cheese, cottage cheese or houmous, and followed by an apple.

Medical advice is to eat at least five portions of fresh fruit or vegetables every day, and I try to follow this, for it seems to be effective in staving off various ills, from strokes to cancer. After apple juice for breakfast plus fruit on my porridge or cereal, multi-veg salad and apple for lunch, I reckon I have taken in my five portions, although I still like to have some vegetables in the evening.

Recently, I visited my optician for a routine eye test and was surprised to be given dietary advice there too. Apparently, I am likely to suffer from loss of vision by age-related macular degeneration, or AMD, which basically means that my retinas will wear out. This is more likely to happen if I am obese, but I can reduce the risk by eating a rainbow of coloured vegetables and fruit. Green leafy vegetables, such as kale and spinach, and red, yellow and orange vegetables, such as carrots, tomatoes, peppers, beetroot, apricots, corn, oranges and red berries, all contain precursors to vitamin A, and you can absorb it better when the vegetables are cooked. In addition, liver, eggs, butter and cod-liver oil are rich in vitamin A. Also useful is vitamin C, found in most fruit and vegetables, and especially in oranges, lemons and grapefruit – best eaten raw – while vitamin E, also good news, is found in nuts, seeds and wholegrains.

I am fond of a kipper for breakfast.

When I am out filming, my diet gets much worse. We may be staying at a hotel, where fruit is not available for breakfast, and knowing there will be a long day ahead, I am inclined to eat bacon, sausage and egg, although I often choose fish if it is on the menu, which is a much healthier option – either kippers or smoked haddock and poached egg. Lunch is usually grabbed on the run – sometimes sandwiches eaten in the car while travelling from one location to another, but usually whatever we can get in the nearest pub. Occasionally, they have salads, but more often they have steak-and-kidney pie, or fish and chips, with hardly a vegetable in sight.

One glorious exception was the week we filmed in Tuscany. Every day we went into the nearest bar, where often children and dogs ran about between the rickety tables, and elegant men came in, downed a slurp of espresso at the bar and walked out again, but the food was delicious. I particularly remember the white beans, which looked dull but tasted

wonderful. I have tried many times to cook them myself, but have never achieved anything like the same result. I shall have to go back one day and ask for some recipes.

Somewhere along the line I have acquired a taste for Japanese food, especially sushi and sashimi – raw fish. Accordingly, I sometimes find myself, just by chance, at the Yo! Sushi bar at Paddington Station in London.

DRINK

On those rare occasions when I have a heavy lunch, I often have a glass or two of wine, but otherwise I never drink alcohol at lunchtime. I find that even a moderate slurp makes me sleepy and inefficient during the afternoon. At Yorkshire Television, I often went into the bar at lunchtime to play chess, and was slightly shocked to see some of my colleagues drinking three pints of Guinness; if I did that, I should do no work at all afterwards. These days I usually drink water at home, and in a pub I have tomato juice and soda, which is surprisingly satisfying.

Somewhere between 4.15 and 5 p.m. I like to have a cup of tea, preferably Earl Grey or a mixture of Earl Grey with something stronger. Earl Grey is a blend of black tea with oil of bergamot, which gives it its characteristic flavour. It first came to Britain as a present from a Chinese mandarin to Charles Grey, the Second Earl, and became a family favourite. He asked his tea merchants to match the flavour, and thereafter anyone could go to them and buy Earl Grey's tea. A politician, and prime minister from 1830 until 1834, Grey was a great reformer and forced through the Reform Act of 1832, which put Britain on the road to proper democracy, and the abolition of slavery in the Empire in 1833.

The history of tea is surprisingly complex and shameful. It was first used as a medicine in the mountains of western China perhaps five thousand years ago, and by AD 200 was apparently used as a substitute for wine. It spread to Japan in around 800 and was brought to the West, by Dutch traders, in around 1600. When Charles II, newly returned from exile, married the Portuguese Catherine of Braganza in 1662, she introduced tea drinking to the English court, and it gradually became more popular until in 1840 Anna, Duchess of Bedford celebrated afternoon tea as a fashionable event in society.

Before this time the Chinese had insisted that we bought their tea with cash, and this was becoming ruinously expensive in silver, so we started growing opium in India and selling that to the Chinese in order to pay for the tea. The Chinese authorities tried to stop this trade, but the British Navy blockaded Chinese ports, and so began the opium wars, which culminated in the British seizing Hong Kong. Meanwhile, botanist Sir Joseph Banks reckoned that we should be able to grow our own tea, and tried first in Assam and later in Darjeeling, south India and Ceylon (now Sri Lanka). In 1815 native tea plants were found growing in Assam; so most of the Victorians' tea came from India.

Tea picking in India.

Whichever sort of tea I am drinking, I like it weak with a little milk but no sugar. When I first went to Eton, I used to 'mess' (have tea) with a greedy lad who ate our entire week's sugar ration on his cereal on Monday, so I learned to enjoy tea without sugar, and have continued to do so for the last fifty years.

There is often a lively debate about how to make the perfect cup of tea, whether the water has to be boiling, whether to put the milk in the cup first, and so on. My normal procedure, if I am making a pot of tea for two or more, is to warm the pot with nearly boiling water as the kettle approaches boiling point, pour away the warming water, put in three tea bags and immediately fill the pot with boiling water. Then I leave it to 'mash' for a few minutes. I put the milk into the mugs first, to make sure it is mixed thoroughly into the tea without having to stir it, and then pour the tea. There are two other slightly dubious reasons for putting the milk in first: the boiling tea might crack the cup or glass if it hits it directly, and the first drops of milk could get scalded and therefore slightly caramelised if it is added to the boiling tea.

The worst possible method, unfortunately used in some railway cafés and trains, is to put the milk in the mug, then the tea bag, and then pour on water that is not very hot. The result is horrible because the true flavour can be extracted from the tea bag only by almost boiling water.

The UK Tea Council (see www.tea.co.uk for details) recommend that you should always use freshly drawn boiling water, and that you should boil the water only once, since otherwise there will be less oxygen in the water. They also say that you should measure the tea carefully – use one tea bag or a rounded teaspoon of loose tea for each cup to be served – and that you should allow the tea to brew before pouring. I can't agree

with them about the oxygen. Water almost always contains dissolved air, but this is driven off as soon as the water warms in the kettle; you can see the little bubbles coming out of the solution. The oxygen must all have gone long before the water boils for the first time. They also recommend using freshly drawn water, but warn that in some places the tap water is heavily chlorinated. If this is the case, then it should be a good idea to let the water stand for an hour or two before boiling, since that will allow at least some of the chlorine to escape.

Shamir Shah, managing director of the East India Tea House, says the water (ideally filtered to remove chlorine) should preferably be boiling if you are making black tea, although the taste does not vary much with water temperature between about 85 and 100°C. However, it does not make any difference whether the water has boiled for the second time. For green tea, the water temperature should be lower, since the leaves scorch at 80°C. For white tea, 70°C is hot enough; white tea is the highest-quality hand-picked Japanese tea and used to be served only to Japanese and Russian royalty.

Leaf tea, Shamir Shah claims, tastes better than tea bags, mainly because the leaves are bigger and retain the flavours better; green tea, Assam and Darjeeling leaf teas have a crisper flavour and a less-bitter aftertaste. Tea in bags is CTC quality – cut, torn and curled – which means that much of the material inside will have been oxidised. This also reduces the amount of antioxidants, which convey a number of health benefits.

Shah says it doesn't matter whether you put the milk in first or last; the taste is the same. He puts his milk in last because every cup of tea is different, and by putting the milk in slowly, he can see how strong the brew is and therefore how much milk he needs.

In the UK, 165 million cups of tea are drunk every day, and 96 per cent of them are made with tea bags, which according to legend were invented in 1908 by Thomas Sullivan in New York; 98 per cent of us drink it with milk. Meanwhile, the teapot was invented in China during the Ming Dynasty (1368–1644).

Until last year we used a large pottery teapot, which I warmed with boiling water before putting in the tea bags and making the tea. Then Sue

Both these heaps are of Earl Grey tea, from the same company, but the top one comes from a tea bag.

was given a smart new stainless-steel teapot, of slightly lower internal capacity, and much smaller external volume, because the steel is much thinner than the clay pot. We thought that the tea would cool considerably faster in the stainless-steel pot, because it is so much thinner and because steel is a good conductor of heat. In practice, the opposite is the case. Even without a tea cosy the tea tastes better in the steel pot and stays hot much longer, and with a tea cosy the effect is even clearer.

I wanted to check this out scientifically, and so I plotted a series of cooling curves. This is a really simple procedure and can be most informative. I rinsed the steel teapot with boiling water for fifteen seconds, then emptied it and filled it to the brim with boiling water, inserted a thermometer probe, put on the lid and measured the temperature, first after thirty seconds, then every minute for a few minutes, then every few minutes. I repeated the process with a tea cosy to keep the heat in. Then I did the same with the ceramic pot, filled with the same quantity of water. The cooling curves are shown in the diagram opposite.

Clearly, the tea cosy helps to retain the heat with both pots. With the cosy in place, the rate of cooling is fairly similar for the two, but the temperature of the tea in the ceramic pot is about 5° lower than in the

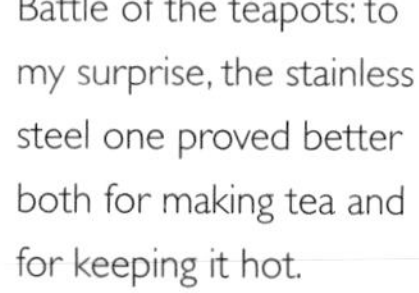

Battle of the teapots: to my surprise, the stainless steel one proved better both for making tea and for keeping it hot.

COOLING CURVES FOR TEAPOTS

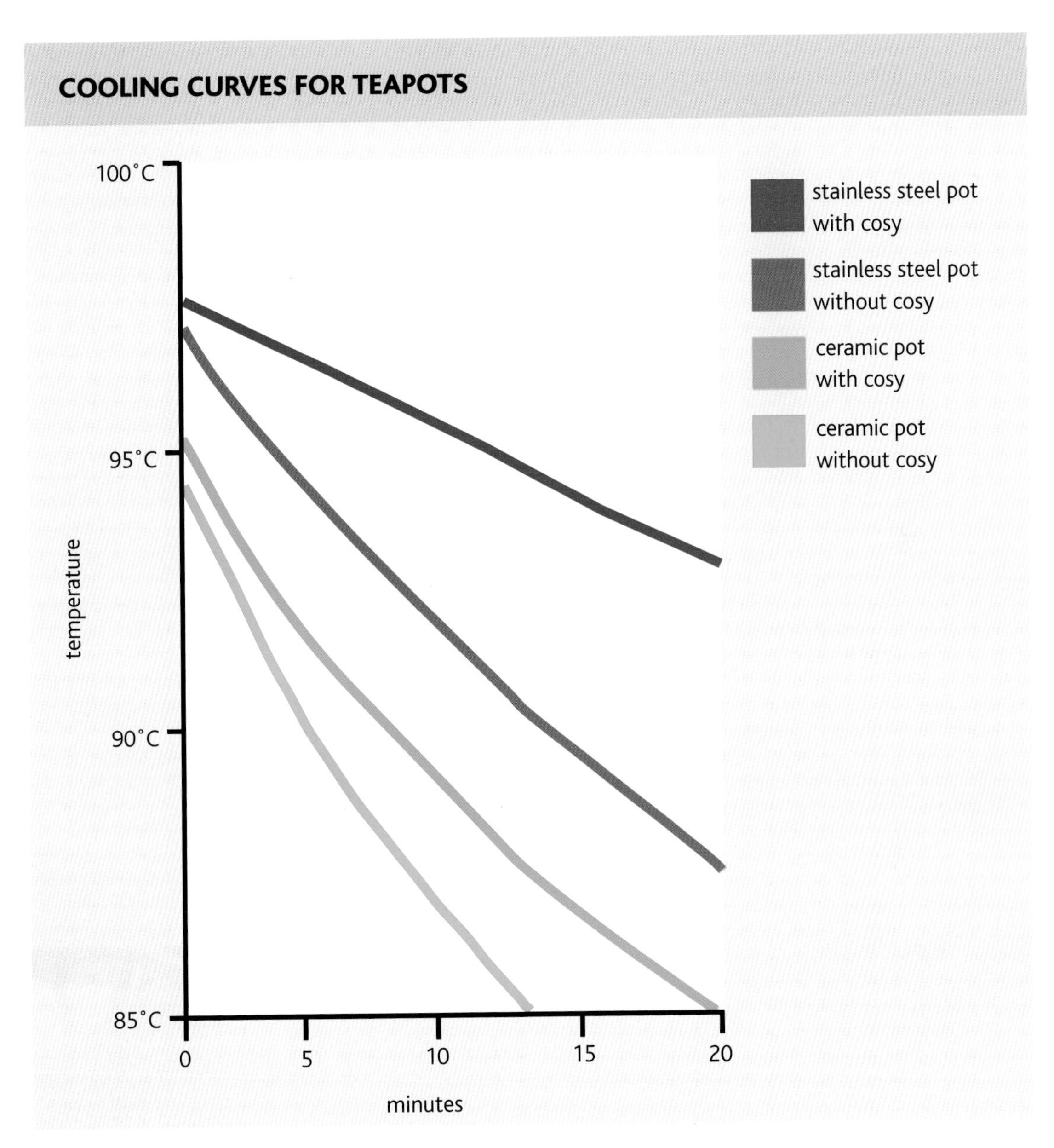

stainless steel. This is why the tea seems to taste better in the stainless steel – the tea leaves are bathed in hotter water – and why it keeps the tea hotter.

Why should this be so? My conclusion is that although the steel pot is thinner and a good conductor of heat, its outer surface is shiny and therefore a poor radiator. Rough, black surfaces are good radiators; smooth, shiny ones are the worst. So it loses heat quite slowly by radiation. Second, the external surface area is lower than that of the ceramic pot, which should reduce heat loss by conduction to the cosy, and by convection. Third, and probably most important, the thermal capacity of the steel pot is much lower – the steel pot is much more easily heated up so rinsing it out for fifteen seconds with boiling water raises its temperature to perhaps 80°C, and when it is filled with tea, the temperature rises well above 95°C. The ceramic pot, on the other hand, needs masses of heat to raise its temperature, and a quick swill probably does not get the pot above 50°C; so the tea hardly reaches 95°C. Even when I filled the pot with boiling water twice before putting the tea in, I was unable to raise the temperature of the tea to 96°C. My conclusion is that teapots should be made of stainless steel or silver, to retain the heat.

Mind you, stainless-steel teapots are not always ideal. I have discovered that in many cafés the steel pots dribble tea all over the tablecloth. Then one day in a small hardware shop I noticed two shelves of stainless-steel teapots for sale. The top shelf had a sign that read, 'TEAPOTS £1.00.' The next shelf said, 'TEAPOTS (POURING) £1.50.'

LECTURING

I gave my first lectures when I was a graduate student at York, in 1969, and I remember being scared beforehand. My first major lecture was to 200 chemistry students in Edmonton, Alberta, at 8 a.m. in mid-winter. I was terrified that my car would not start, so I set off from home about two hours too early, but everything went fine. Since I have appeared on television, I am asked to give talks all over the place; indeed, I get about three invitations every day, and most of them are for evening talks. Science Week, in mid-March, is a particularly busy time; I may have a talk to give every day. Fortunately, they don't all have to be different, but unfortunately, they are all over the country, and I spend more hours on trains than I care to think about. For these lectures I used to use 35-millimetre slides to illustrate what I was talking about, but now most lecture theatres have digital projectors, and I run similar slides on PowerPoint.

Giving talks is fun, especially to audiences of a few hundred, and ideally with hugely mixed audiences including children and elderly people. Small kids with their grandparents are particularly receptive, and the children ask such splendid questions. After one talk at the Royal

Institution a small girl asked, 'Have you ever broken anything expensive?' and I wondered what she had done that morning. At another talk a little boy asked whether worms sleep; I had to admit that I don't know, but since they have distributed nervous systems rather than brains, which need to shut down for a rest, the answer is probably no. And more recently I was asked why old people don't have so many brains; I was tempted to say I was too old to remember, but in fact I tried to explain a little of why memory deteriorates. Various diseases can affect the brain, and memory is often diminished by strokes, but even without disease or strokes memory gets worse in old age because the fatty insulation of the neurons begins to break down, with the result that there are some wrong connections and short circuits (see page 174).

The worst speakers, in my experience, are those who read the entire lecture from a script. For one thing, the written word almost always sounds stilted when read aloud, and for another, the speaker cannot make eye contact with the audience while reading; it looks as though the talk is being addressed to the lectern. Often, too, such lectures are mumbled in a monotone. I try to make mine informal, almost a conversation with the audience, and the more I can engage them and interact with them, the happier I am. I can then gauge their mood and divert to particular anecdotes if I think they will be appreciated. That is why questions are such fun; they are the most interactive and unpredictable part of the event.

I enjoy talking about science and photography, and my love for both of them. I am often asked to give talks about aspects of the history of science, and I am delighted to be able to use some of the mountains of material we collected during the research for *Local Heroes*. This year, 2006, is the 200th anniversary of the birth of flamboyant engineer Isambard Kingdom Brunel, and I shall give several talks about him and his Victorian rivals, probably ending with the words that his colleague Daniel Gooch wrote in his diary when Brunel died, that I.K.B. was 'bold in his plans but right. The commercial world thought him extravagant, but altho' he was so, great things are not done by those who sit down and count the cost of every thought and act.'

Isambard Kingdom Brunel, perhaps the greatest of the Victorian engineers.

6 After Work

SUNDOWNER

At lunchtime I don't often drink alcohol, but in the evening I do like to have a drink, and I am not fussy; once the sun has crossed the yardarm (6 p.m. in my book), I will drink almost anything. I generally have whisky and soda or ginger ale, but Sue drinks gin and tonic, and I often join her.

Mixing gin and tonic seems simple enough, but I have heard serious arguments about how it should be done. The basic ingredients are gin (one finger's width in the glass), ice cubes, a slice of lemon or lime, and tonic, probably enough to fill the glass. One camp asserts that the correct procedure is first to pour in the gin to the required depth, then add the slice of lime or lemon, crushing or twisting it a little as you put it in, so as to mix some citrus juice with the gin. Add a couple of ice cubes, then fill the glass with tonic; by pouring it in, you will mix all the liquid ingredients and make the drink homogeneous – i.e. the same all the way through.

The opposite method is to fill the glass with crushed ice, add the slice of lemon and pour the tonic over it, and finally pour the gin carefully on top. This means that everything will be well cooled by the mass of ice. The lime or lemon juice will be well distributed, but most of the gin will sit on top of the tonic and will mix only slowly. The advantage is that the first mouthful will be mainly gin and will therefore pack a wallop. The disadvantages are that you don't know how much gin to add, unless you use a separate measure, and the drink gets steadily weaker as the gin concentration goes down.

Then a further problem follows, if you take the tonic (or ginger ale, Coke or other fizzy drink) from a plastic bottle. You use only a little and want to keep the rest as fizzy as possible for the next day. What is the best method? Should you just screw the top on quickly? Blow into the bottle to maximise the carbon dioxide concentration and then screw the top on? I squeeze the bottle so that the liquid comes right up to the top, then screw the top on to the squeezed bottle and put it in the fridge. Keeping it cold must be a good thing, since gases are more soluble in cold liquids than in hot ones, so the carbon dioxide is less likely to bubble out if the tonic is cold. Squeezing the bottle minimises the volume of gas in the bottle, and I think this will minimise the amount of carbon dioxide that will bubble out of the solution. However, several people disagree with my theory, and I have yet to devise the definitive test to find out whether I am right. Anyone who has a better idea is most welcome to write and tell me about it.

ICE IN THE DRINK

Recently, I was given a present of two little brightly coloured plastic fish that were supposed to be cooled in the freezer and then dropped into the lunchtime ginger beer or the medicinal gin and tonic to cool it down. I tried them out, but was disappointed. Not only did they look silly, and therefore spoil the pure appearance of the drink, they also failed miserably to keep it cool.

Temperature is a measure of how quickly the molecules of material are moving about and vibrating, so the molecules in hot water are zooming about and wriggling much more quickly than those in cold water, which in turn are more active than those in ice. When you bring together something hot and something cold, some of the extra energy of the hot molecules is transferred to the cold ones; they get stirred up and kicked into action. The opposite cannot happen; the sluggish cold molecules cannot make the hot ones move faster; they simply don't have the energy. That is why heat always flows from a hot thing to a cold one (this is the zeroth law of thermodynamics), and why heat will flow from the warm liquid to the cold ice cubes.

What is the perfect recipe for gin and tonic?

In practice, ice cubes cool your drink by two slightly different processes. First, they are cold – they come out of the freezer at perhaps –10°C, while the liquid may be at, say, 20°C – and therefore heat will flow from the drink into the ice cubes, so that the drink cools down. Second, and much more important, is the effect of the latent heat of melting. When the cold ice is put into the warm drink, its temperature rises steadily to 0°C. At that point it begins to melt. Then the temperature of the ice remains 0°C, but the process of melting extracts heat from the liquid. Because the frozen molecules of H_2O are tightly bound together in the solid ice, lots of heat is needed to break them away and release them into the liquid. In fact, to melt a gram of ice takes 80 calories, which is enough to cool down 80 grams of water by 1°C. Therefore as it melts, a single ice cube (about 15 grams) will cool an average drink of 200 grams by 6°C, and four ice cubes will get the whole drink down from room temperature (in Britain) to freezing point.

Joseph Black, who discovered latent heat, and therefore why ice cools your drink.

The reason why the plastic fish don't work is probably because they contain only a little liquid (which may be water; few other liquids have such a large latent heat of melting), and the plastic container merely impedes the flow of heat from the liquid to the material inside.

The phenomenon of latent heat was first discovered by an elegant Scottish scientist, Joseph Black. Some of his pupils at Edinburgh University were the sons of whisky distillers, and they wanted to know why they had to spend so much money distilling the whisky; masses of fuel was needed to provide the heat, and then they had to cool it all down again afterwards. You can see the effect for yourself: put a pan of water on the stove, light the gas and measure the temperature. The higher you turn up the gas, the faster the temperature goes up – until it gets to 100°C. At this point, the water begins to boil, and however much you turn up the gas, the temperature will not go over 100°C. So what is happening to all the heat you are putting into the pan? The answer, Black realised, is that the process of turning liquid into vapour requires heat; energy is needed to get those molecules of H_2O moving fast enough to escape from the surface as steam. Because this heat energy seems to be going nowhere, Black called it latent heat, meaning hidden heat; this is the latent heat of evaporation.

Black almost certainly explained his ideas on latent heat to his young friend James Watt, and as a result, I guess, Watt saw why the existing Newcomen steam engines were so inefficient, and conceived the idea of installing a separate condenser. The Newcomen engine had an enormous cast-iron cylinder that had to be first filled with steam, to allow the piston to rise, and then cooled with a jet of cold water, so that the steam condensed to make a vacuum and atmospheric pressure forced the piston down. This meant that on every stroke the entire cylinder had to be heated above 100°C before it could be filled with steam, and then cooled below 100°C so that the steam would condense. Watt realised that the engine would be much more efficient if the working cylinder could be kept hot, at, say, 110°C, and the steam condensed somewhere else – in a separate condenser. All it needed was a pipe coming off and leading into a cooled vessel. At the right moment a tap could be opened to allow the steam to escape into this pipe, and it would all condense outside.

Technically, this was extremely difficult to achieve – it took Watt ten years to get his first engine working – but the idea came to him in a flash one Sunday in May 1765, while he was walking on Glasgow Green. He later said, 'I had not walked further than the golf house when the whole thing was arranged in my mind.' This was Watt's eureka moment, and why we have all heard of him, for he went on to make the best steam engines in the world.

Joseph Black was a delightful man, always beautifully dressed, and with impeccable manners. He died in the evening of 6 December 1799, surrounded by his family. He had just eaten his supper of a little bread and some prunes, and had taken a drink of milk from a bowl; he leaned

The industrial revolution began in 1712 with the Newcomen engine, but it occupied a large building and was woefully inefficient.

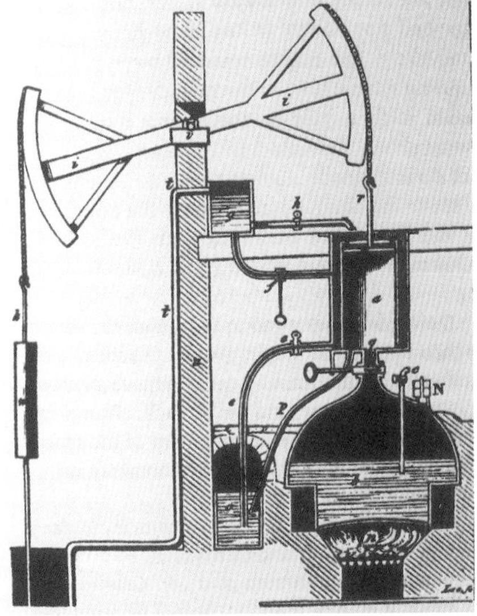

forward, put the bowl down and died, without spilling a drop. Those standing round said it was almost as though he had deliberately done an experiment to see whether he could die as precisely and as peacefully as he had lived.

DINNER

When Sue and I are both at home, we usually have simple soup for dinner. One of us will make a large batch, using vegetables from the garden when they are available, and this lasts about three days. Soup made mainly from home-grown butternut squash is particularly satisfying, but I am happy to include potatoes, leeks, carrots, turnips or almost any winter vegetable. A summer favourite is tomato egg-flower soup, made by simmering chopped tomatoes and tofu chunks in stock with soy sauce and then stirring in a thin stream of egg beaten with toasted sesame oil, the whole dish garnished with chopped spring onions.

Once or twice a week, however, and whenever we have friends visiting, I am allowed to cook something more substantial. This is nearly always fish – we rarely eat any meat – and may be anything from grilled herring to baked trout with fennel. Oily fish have great strong flavours and are also loaded with omega-3 unsaturated fatty acids. There is some controversy about these compounds, but they are said to be excellent brain food – important for repairing damage to the insulation of the neurons. Our brains are made of millions of neurons, which are tremendously long, thin cells that are interconnected with enormous complexity and carry electrical signals to one another. The cells are insulated with fatty sheaths, just as electrical wires in the home are insulated with plastic coating. When the insulation breaks down, we are liable to get wrong connections or short circuits in the brain, which is why mental deterioration often accompanies old age.

Neurons, or nerve cells in the brain.

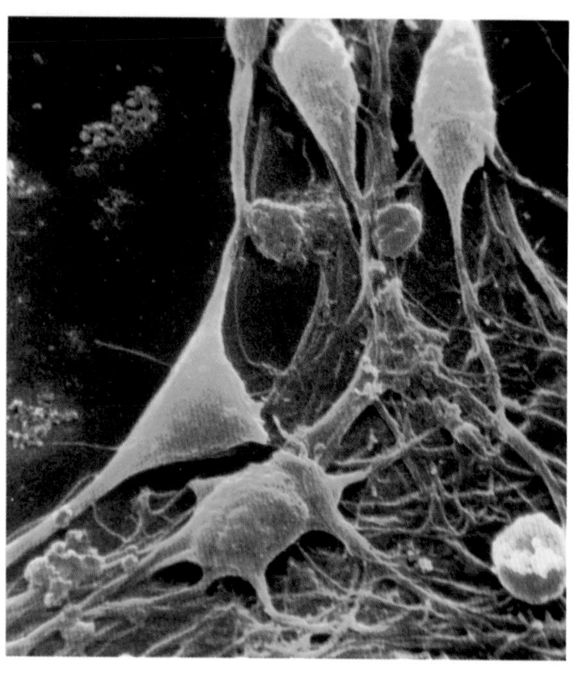

I enjoy cooking, but I have to keep it simple, and I like to work from recipes, so that I (think I) know what I am doing. Even then, I sometimes forget one of the ingredients, so my meals are a bit hit-and-miss, but I do enjoy making them, and even more so if the guests eat the lot. Occasionally, I invent my own recipes, and am delighted when they work; a recent successful experiment was to have plain slightly stewed rhubarb with grilled mackerel.

Cooking was one of the most important primitive inventions. It probably started when someone threw some food remains into the campfire and noticed the delicious smell. Then they must have tried briefly putting the food on or in the fire before eating it, and found that this improved the taste. Grilling or barbecuing meat is easy, because of the fat in the meat. All you need is a sharp stick to impale the meat on and a fire to hold it over. Porridge needs another level of technology – the cooking pot. Once people had bronze (about 2000 BC), they could fashion it into saucepans, which must have been highly prized items. Long before that, however, people had made clay pots, and some of these were probably tough enough to be usable as cooking pots. A thick earthenware pot placed directly on a hot fire might crack, but a thin glazed pot probably has a better chance, especially if it is placed on a stone beside the fire, or if it is filled with the water and other ingredients to be cooked and very hot stones from the fire are dropped gently inside. This will bring the water to the boil without heating the pot directly, and clay pots should be able to stand water boiling at 100°C.

An earthenware cooking pot discovered in the wreck of a fourteenth-century ship in the South China Sea.

Cooking is important for several reasons. First, it improves the taste of many foods, especially meat. Second, it greatly extends the range of possible foods by making edible many things that cannot be eaten raw – converting grain to porridge is a good example. More important, it destroys toxins in many foods, from beans and rhubarb to cassava or manioc, which is a staple food in many parts of the world but seriously poisonous when raw. Third, cooking kills vast numbers of bacteria and parasites (worms and eggs, for example) that would otherwise be likely to infect the eaters. Fourth, it tenderises many foods, and therefore makes them much easier to eat and to digest; carrots, for one, can be eaten raw, but are actually better for you if they are slightly cooked.

The chemistry of cooking is complex. Roasting, grilling and barbecuing slightly burn the outside of the material, by reaction with oxygen in the air. This caramelisation strips off hydrogen atoms from the organic chemicals in the food and leaves an increasing concentration of carbon, which makes the surface steadily darker, until if you go too far it is thoroughly charred; the black charring is mainly carbon in the form of soot.

Inside, the effect of the heat penetrating from the surface is to denature the proteins. This is an important process, which involves breaking up the long protein molecules or polypeptides into shorter ones called oligopeptides. This makes them easier for us to digest, since denaturing is a vital part of the digestive process. Because the heat has to be conducted in from the outside, the inside of any piece of meat or fish always takes longer to cook than the outside, and being an impatient and incompetent cook, I quite often find that I have taken fish from the grill or the oven too soon; when I cut it open, I find the middle is still raw.

Any cooking with water in a pot is partly to do with denaturing proteins, partly to do with hydrolysis – breaking down large molecules by reaction with water – and partly to do with dissolving various things in the water to change their taste and texture. Pressure cookers work in the same way, but more quickly, because under pressure the water boils at a higher temperature.

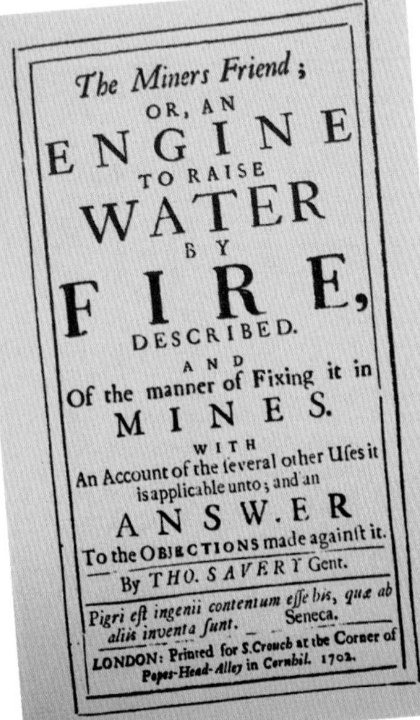

The Miners Friend;
OR, AN
ENGINE
TO RAISE
WATER
BY
FIRE,
DESCRIBED.
AND
Of the manner of Fixing it in
MINES.
WITH
An Account of the ſeveral other Uſes it
is applicable unto; and an
ANSWER
To the OBJECTIONS made againſt it.
By THO. SAVERY Gent.

Pigri eſt ingenii contentum eſſe his, quæ ab aliis inventa ſunt. Seneca.

LONDON: Printed for S. Crouch at the Corner of *Popes-Head-Alley* in *Cornhil.* 1702.

Thomas Savery's 1702 book *The Miner's Friend, Or, An Engine to Raise Water by Fire*, but his steam engine was not a success.

The pressure cooker was invented by an ingenious French scientist, Denis Papin, who moved from Paris to London, and in 1682 demonstrated his 'New Digester or Engine for Softening Bones' to the Royal Society. What is more, on 12 April that year he laid on a philosophical dinner for the president and fellows, after cooking the whole meal in his digester. Sadly, we have no record of the menu, but we do know that the dinner was a great success, since John Evelyn the diarist was among the guests and wrote warmly of the occasion. We also know that Papin was enthusiastic about using his digester to cook 'old rabbet', which he considered 'ordinarily but a pitiful sort of meat', because it made it taste young and tender, and turned the bones to excellent 'gelly'. Today's pressure cookers raise the pressure to about 1.5 or 2 atmospheres, while Papin's apparently reached a terrifying 6 atmospheres.

Papin went on to build both low-pressure (atmospheric) and high-pressure steam engines, and may well have been the spur that drove Thomas Savery and Thomas Newcomen. Savery took out a patent in 1698 for 'raising water by the impellant force of fire', but Thomas Newcomen, an ironmonger from Dartmouth, was the first person to build a really useful working steam engine, at Dudley in the West Midlands, in 1712. The Newcomen engine was woefully inefficient and was therefore useful only for draining coal mines, where the fuel was free. Nevertheless, this was the machine that started the Industrial Revolution, for after improvements by James Watt, sixty years later, steam engines began to be used to drive industrial machinery. I am entertained by the thought that the Industrial Revolution began with the pressure cooker.

James Watt trying to work out how to make an efficient steam engine.

Sometimes I use a microwave oven, which is convenient because it is quick and easy, and it works a treat, for example, with white fish thinly smeared with a little herby butter. Microwaves cook food in a different way from normal cookers. Microwaves are high-frequency radio waves, with a wavelength of around 12 centimetres. They can conveniently be channelled down metal tubes or waveguides with a rectangular cross-section of around 6 centimetres by 3 centimetres. Such tubes guide the

My microwave oven, after stewing some fruit.

A thermogram of a microwave oven after use. The hottest thing is the food (white), then the bowl, then the inside of the oven.

microwaves into your oven, where they bounce around between the walls and the wire mesh on the inside of the door; they can't escape through the door because the holes in it are much smaller than the wavelength. As they bounce around the oven, they pass through any food inside. The energy of the microwaves is absorbed by water, which is rapidly heated as the radio waves make the water molecules spin, and so the microwave energy is converted to heat energy. This rapidly cooks the food, and because the waves pass right through it, the food is cooked not just from the outside, but on the inside at the same time. So unlike my fish in the normal oven, the food cooked in a microwave oven is much more likely to be cooked right through.

One slight problem is that the distribution of the waves inside the oven may not be entirely homogeneous, and so rotating the food is a good idea, since then the waves will come at it from all directions. Also, the microwaves would rapidly cook your hand, or any other part of your anatomy exposed to them, so the microwave oven has to be designed so that the waves cannot escape, and so that they are switched off when you open the door.

One of the few tiresome things about eating at home is that the dishes need washing afterwards. I have great admiration for the crazy cyclist Josie Dew, who writes wonderful books (try *The Wind in My Wheels*) and has been a professional cook, but travels extremely light; she says you can eat anything from a saucepan with a spoon.

Washing the dishes by hand is a bore at any time, and a chore after dinner, especially if the dinner was elaborate and there were more than two or three people. Rubber gloves are uncomfortable; strong detergents attack the skin; and water hot enough to clean the crockery is also hot enough to be painful on the hands. As a research chemist, I learned that a flask full of liquid at 60°C is just too hot to hold in the hand – a useful lesson, if expensive in flasks.

Dishwashers really do make life easier. They mix strong detergent with hot water (at about 60°C or 140°F) and squirt it vigorously at the crockery and cutlery, then repeat the process with rinsing water and may also heat the air to help everything dry. You have to learn to pack the crockery in carefully, so that bits aren't shielded from the jets, or they don't get clean, but in general I am in favour of the dishwasher.

The modern dishwasher saves time and effort, and is probably more hygienic and more energy efficient than washing dishes by hand.

The first one was patented in 1850 by Joel Houghton. It held the dishes in a wooden rack and splashed water on them when you turned a handle. My guess is that it was less effective than washing by hand, and possibly slower too, but Houghton was the pioneer. In 1885 M. Eugene Daugin constructed for a Paris restaurant a wonderful machine with eight revolving artificial hands that each held a dish, dipped them in hot soapy water and then in clean rinsing water. Finally, a pair of revolving brushes completed the process, and the dishes were left to dry in a rack.

Josephine Cochran, American socialite, patented the first useful automatic dishwasher in 1886.

The credit for the first effective automatic dishwasher, however, goes rather surprisingly to an American socialite from the Midwest called Josephine Cochran, or Cochrane – she seems to have added the final 'e' because it looked grander. She was a great entertainer in Shelbyville, Illinois, and was sad to see that her beautiful inherited china was getting chipped by the servants, who were careless while they were washing up. Her first remedy was to do the washing up herself, but she had inherited an engineering streak, and she went out to the shed in her backyard and devised a machine to do the job. All the plates and cups and cutlery were held on racks inside, and hot soapy water was sprayed vigorously on them when she turned a handle. Later models were driven by pedal power and apparently by steam.

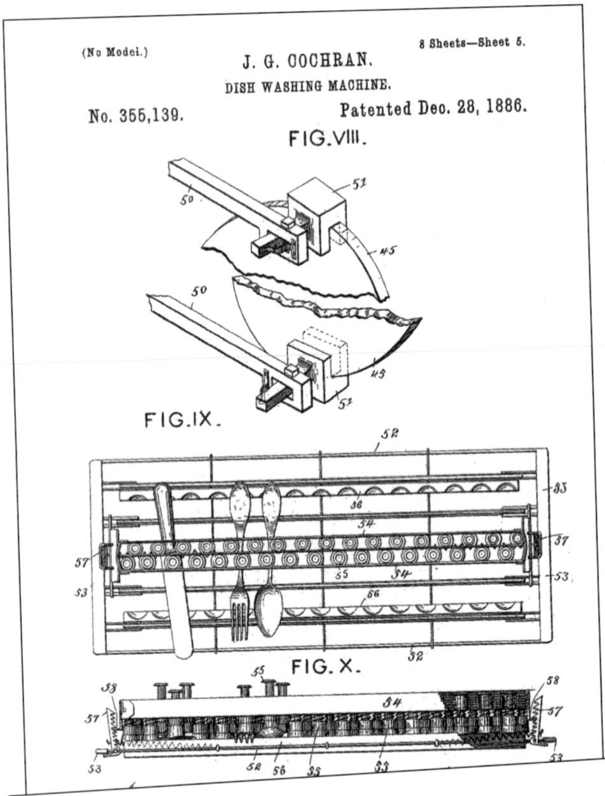

(No Model.) J. G. COCHRAN. 8 Sheets—Sheet 5.
DISH WASHING MACHINE.
No. 355,139. Patented Dec. 28, 1886.
FIG.VIII.
FIG.IX.
FIG.X.

UNITED STATES PATENT OFFICE.

JOSEPHINE G. COCHRAN, OF SHELBYVILLE, ILLINOIS.

DISH-WASHING MACHINE.

SPECIFICATION forming part of Letters Patent No. 355,139, dated December 28, 1886.

Application filed December 31, 1885. Serial No. 187,276. (No model.)

To all whom it may concern:

Be it known that I, JOSEPHINE G. COCHRAN, ... residing at

other, the soap-suds and clear hot water, when respectively used, being each returned to its appropriate force-pump and used over and ... needs renewing.

She patented her invention in 1886 and exhibited it at the 1893 World's Columbian Exposition in Chicago, otherwise known as the World's Fair, but unfortunately, only a few hotels were interested. As a domestic appliance, the dishwasher did not really catch on until about fifty years later. Just like Sandy Bain (see page 127), she was decades ahead of her time.

ARTIFICIAL LIGHT

We take it for granted that we can light our homes in the evenings, that we can read and play games, and see what we are eating, but this is relatively recent (see page 18). Two thousand years ago the Romans used oil lamps or candles, and these remained the only effective sources of artificial light until the nineteenth century. Oil lamps produced poor yellow light and lots of smell and sooty smoke; candles were rather better, but expensive and only for the rich. Both represented a serious fire hazard – the last fire at Hampton Court Palace, in 1986, was caused by an overturned candle. Several rooms were destroyed, including the King's Privy Chamber, and it took six years to restore them to what they would have been like in 1700.

Candle lighting in Paris, 1667.

What's more, both oil lamps and candles needed to be lit, which was difficult with flint and tinder, and even worse if you had to rub two sticks together. In the early 1800s there were fiercely dangerous 'Promethean matches' – glass vials of concentrated sulphuric acid wrapped in paper impregnated with potassium chlorate, sugar and sulphur to make it burn easily. Crack the glass with pliers – or with your teeth if you were in a macho mood – and the acid would escape, setting fire to the paper. Not to be recommended for lighting your bedside candle. We have to assume that during the winter most people got up in the dark.

There was a great leap forward in 1826, when John Walker of Stockton-on-Tees invented the friction match. He was the proprietor of a chemist's shop on the high street, and had a wide enthusiasm for science in general, and chemistry in particular. He was experimenting with various combinations of chemicals on wooden splints and found one mixture that caught fire when he scraped it on the bricks by his chimney, and in turn set fire to the wooden splint. He refined the chemistry, began making matches and made his first sale, to a solicitor, Mr Hixon, on 7 April 1827. They came to be called Friction Lights, and he charged a shilling (5p) for a tin of eighty-four.

He refused to patent his invention, even though he was advised to do so by many people, including Michael Faraday; he said, 'I doubt not it will be a benefit to the public. Let them have it. I shall always be able to obtain sufficient for myself.' Walker's has been described as a perfect invention. There had long been a need for a simple ignition device, and he produced one, which has scarcely changed since. Today's matches contain compounds of phosphorus and sulphur, and – just like Walker's – ignite as a result of the heat generated by friction.

A new source of light that came into use during the Victorian period was coal gas, which had been discovered about 1690 by John Clayton, rector of Crofton, near Wakefield in Yorkshire. He wrote a wonderful letter to Robert Boyle at the Royal Society in London, describing a ditch within 2 miles of Wigan 'wherein the water would seemingly burn like

The first report of coal gas, about 1690.

V. *An Experiment concerning the* Spirit of Coals, *being part of a Letter to the Hon.* Rob. Boyle, *Efq; from the late Rev.* John Clayton, *D. D. communicated by the Right Rev. Father in God* Robert *Lord Bifhop of* Corke *to the Right Hon.* John *Earl of* Egmont, *F. R. S.*

——HAving feen a Ditch within two Miles from *Wigan* in *Lancafhire*, wherein the Water would feemingly burn like Brandy, the Flame of which was fo fierce, that feveral Strangers have boiled Eggs over it; the People thereabouts indeed affirm, that about 30 Years ago it would have boiled a Piece of Beef; and that whereas much Rain formerly made it burn much fiercer, now after Rain it

Brandy, the Flame of which was so fierce that several strangers have boiled eggs over it'. He drained the ditch and lowered a lighted candle into it, and 'the Air catched Fire, and continued burning'. Then he dug down and found some 'shelly coal', which he took home, heated in a retort on an open fire and made an inflammable gas, which he called Spirit of Coal and stored in bladders to amuse his friends.

Coal gas, produced by heating coal, is a mixture of several gases, mainly hydrogen, methane and carbon monoxide, but also includes small amounts of ammonia, hydrogen sulphide and some other hydrocarbons and sulphur compounds. The first three, hydrogen, methane and carbon monoxide, burn well, but would produce hardly any light if they were pure – just a pale-blue flame. The impurities, especially the heavier hydrocarbons, burn inefficiently to produce particles of soot, which glow to make the light, but then settle on the ceiling in black smudges. Carbon monoxide is poisonous, while hydrogen sulphide and other sulphur compounds are not only poisonous but also extremely smelly. So this intriguing and useful cocktail of gases would have some trouble passing today's health and safety standards.

In the 1790s an inventive Scot, William Murdoch, lit on the idea of using coal gas for illumination and succeeded in lighting part of his house at Redruth. He was working at the time for the Birmingham-based engineering firm Boulton & Watt, who had sent Murdoch down to Cornwall to look after their interests there. Unfortunately, Matthew Boulton was keen neither on his employees inventing things nor on taking on new challenges, so even though in 1802 Murdoch lit the entire Soho Foundry at Birmingham with gaslight, and four years later went on to install 900 burners to illuminate a Lancashire cotton mill, Boulton refused to patent the idea – he said they had enough patents already – and Murdoch made no money from his invention.

Nevertheless, the idea took hold. By 1816 there were 26 miles of gas mains in London, and many streets and public buildings had gaslight. It still wasn't much use for houses, though, since the flickering flames from 'fishtail' burners scarcely gave enough light to read by, while the gas was poisonous and the fumes were smelly and dirty. In spite of these problems, gas lighting gradually transformed Victorian Britain. In winter the working day was lengthened, the streets were made much safer at night, and people of all classes were able to read, and therefore acquire an

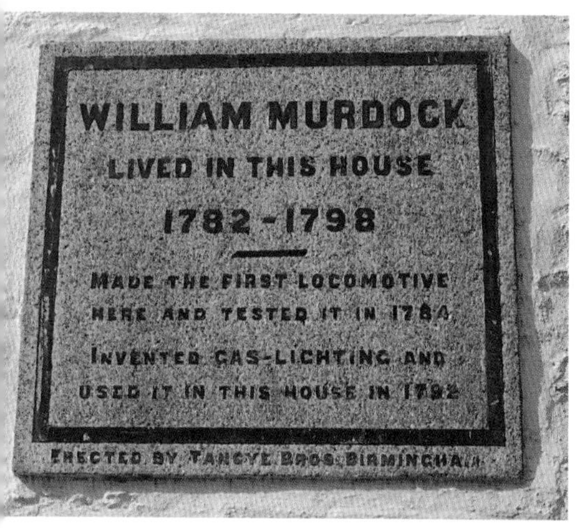

A plaque on the wall of the house in Redruth occupied by William Murdoch (or Murdock).

education. Dinner, which had normally been taken at, say, 3 p.m., gradually slid into the evening.

Steadily, the quality of the gaslight improved; the invention of the Bunsen burner in 1855 allowed gas to burn at a much higher temperature, and then the improved gas mantles made from the oxides of the exotic metals thorium and caesium allowed a much brighter, cleaner light. Meanwhile, the smell had been much reduced by extracting the sulphur compounds from the crude gas. From about 1890, therefore, coal gas gave a good steady, bright light – but by then it had to compete with electricity (see page 21).

Now we all illuminate our homes with electricity, but for a soft romantic evening glim, Sue and I sometimes turn back the clock and light candles, whose flickering glow is plenty to eat by, especially with a glass of wine at hand. During dinner we often share a bottle of wine. Sue continually tries to educate me about which grapes and which years I should enjoy, but the fact is that I enjoy almost any wine that is put in front of me – and candlelight is hopeless if I want to read the label on the bottle.

THE REMAINS OF THE DAY

I enjoy making bread, usually using a mixture of wholemeal and granary flour, and this is one of the few things for which I have the confidence to forget the recipe and proceed from memory and instinct. My bread occasionally emerges a bit heavy and underdone, but is almost always popular enough to disappear rapidly. Doing the first stage in the evening and leaving the dough to rise overnight often works well.

Sometimes, after dinner, I take a wee dram of single malt, and my favourite is Laphroaig cask strength; I was brought back to single malts by reading Iain Banks's entertaining book *Raw Spirit*. I rarely go out to drink – I have never been into any of my local pubs. I prefer to drink at home, with friends or at dinners or receptions, of which I get invited to many.

After dinner at home we occasionally watch television, though that is scarcely more than twice a week. We look out for such satirical programmes as *Have I Got News for You* and *Dead Ringers*, and we like to see science programmes, but sadly there aren't many on television these days – too many programmes on cooking, house improvements, DIY and gardening. When I started work at Yorkshire Television, in 1977, we made a wacky science programme called *Don't Ask Me!*, with Magnus Pyke, David Bellamy and Miriam Stoppard. The series ran for 26 weeks, all through the summer. Meanwhile the BBC was making *Tomorrow's World, QED, Antenna* and *Horizon*. Now those series have all gone, apart from *Horizon*, and although there are a few science programmes on BBC2 and other channels, there are nothing like as many as there were 30 years ago.

Recently I read in the newspaper that the original hope for television from John Logie Baird, who invented it, was that it would be a great medium for international education and world peace. As his son pointed out in this article, John Logie would have hated *Big Brother*, and *I'm A*

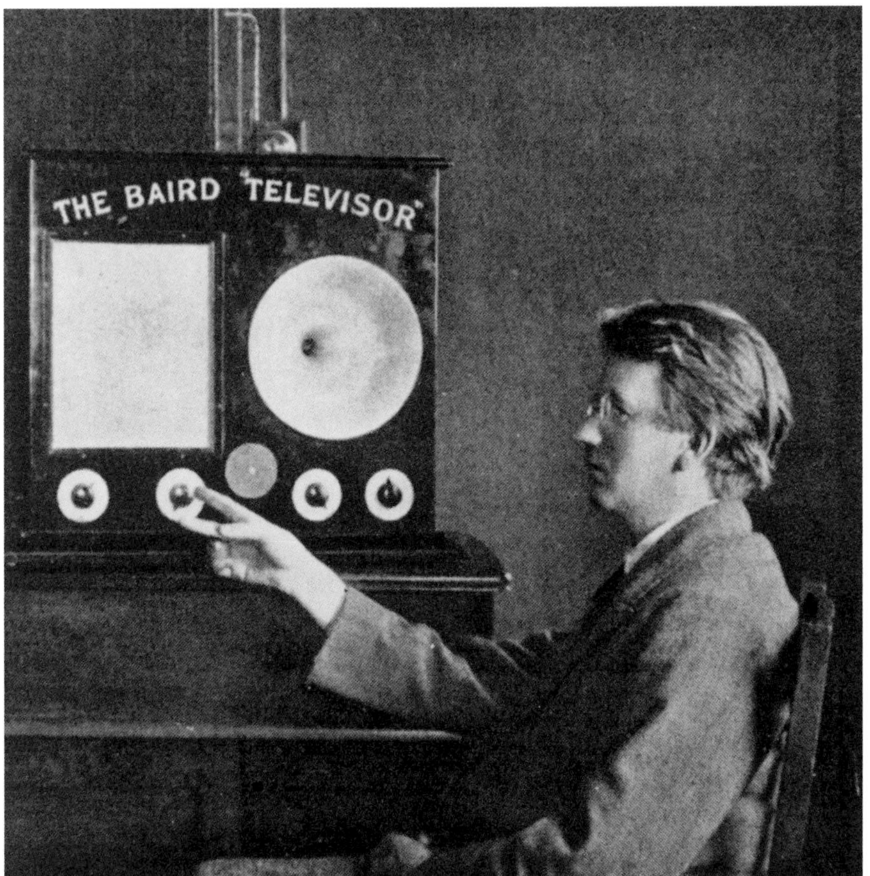

John Logie Baird, inventor of television.

Celebrity... Get Me Out Of Here. John Logie Baird was a remarkable man. A sickly Scot, he was born in 1888 and grew up in Helensburgh, near Glasgow, and was always inventive. As a lad, he rigged up a telephone system between his house and those of his friends, but unfortunately one of the wires dangled too low over the road and lifted a cab-man right out of his seat.

By the age of 13 he was obsessed by the idea of television – the word was coined in 1900. His dad wanted him to go into the church, but he enrolled in the Royal Glasgow College of Technology, where a fellow student was John Reith, who was to become the first Director-General of the BBC. They disliked one another. Baird was unfit for service when the Great War broke out in 1914, but worked on maintaining Glasgow's electricity supply, and after the war managed to make £200 a week by selling Baird Patent Undersocks from a barrow, using fictitious testimonials from satisfied customers. At the age of 22 he decided to spend all his time trying to get television to work, and eventually, in spite

A family watching a Baird television, 1930.

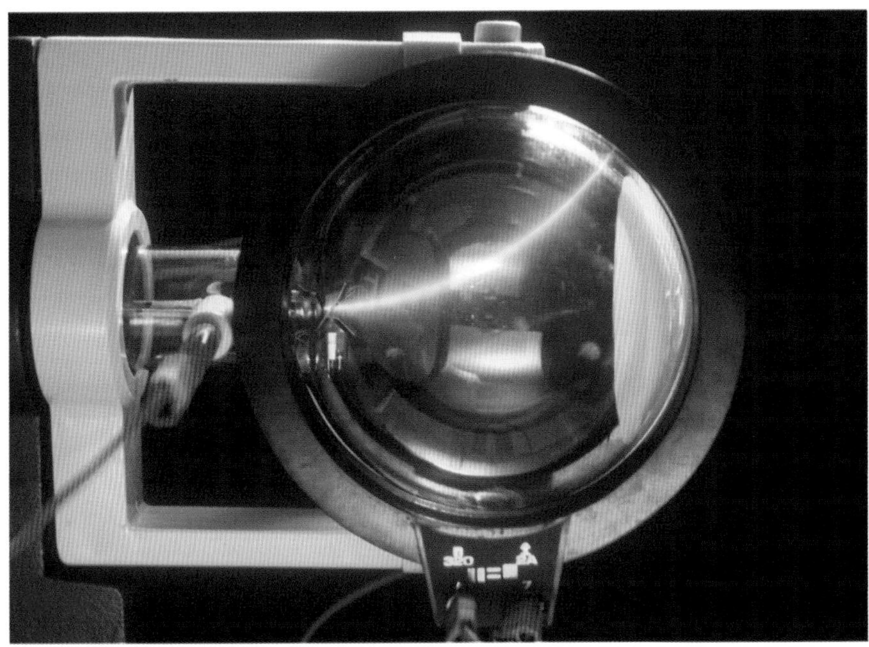

The cathode ray tube was a vital part of television sets for most of the twentieth century.

of desperate poverty, he had his first public showing in March 1925 in Selfridges, the London department store. On 3 June 1931 Baird televised the Derby from Epsom, and arranged for a receiver to be set up in 10 Downing Street. The Prime Minister, Ramsey MacDonald, wrote to Baird: 'When I look at the transmissions I feel that the most wonderful miracle is being done under my eye... You have put something in my room which will never let me forget how strange the world is – and how unknown.' At least I can be sure Baird and MacDonald would both have enjoyed watching Sir David Attenborough, my great TV hero.

Sue and I sometimes play Scrabble or another word game. Most often I read for a few minutes before settling down. I usually have one heavy book on the go – at the moment Jared Diamond's *Collapse* – and one easy read – at the moment Nevil Shute's *Pied Piper*. I also use this time to read a few of the many periodicals and magazines that come through the letter box – including *The Week*, three cycling magazines, two science magazines, two photographic magazines and all sorts of glossy publications from a dozen universities, the Royal Society of Arts, the Royal Institution, the Institution of Lighting Engineers, the Arthur Ransome Society, the British Toilet Association, and the Association of Lighthouse Keepers.

GOING TO SLEEP

Unless I am at some social event, I go to bed between nine and ten. I brush my teeth again, put on a T-shirt and some boxer shorts, take a glass of water to the bedside table and lie down. When I was young, I used to have on my bed a bottom sheet, a top sheet and then blankets, perhaps two or three, depending on their thickness and on the temperature, or perhaps one blanket and an eiderdown, which was a quilted covering containing feathers – ideally, the under feathers, or down, of the eider duck. Now the eiderdown has disappeared, and in many cases blankets have disappeared too, replaced by the lightweight but highly insulating duvet, which seems to have migrated across the Channel into Britain from Continental Europe and displaced the native species.

At first I found duvets rather difficult to sleep under. I was used to blankets holding me down, partly because they were heavy, and partly because they were tucked in tight on both sides of the bed. Duvets are light and are not tucked in. I have accepted this now, and don't often kick the duvet off in the night, even though it is not anchored, but still I have one problem: I often find them too hot. In the old days I had the option of throwing off one or more blankets to get the temperature right, but the duvet is either on or off, and all too often I find that off is too cold but on is too hot.

At home, my answer is to have a top sheet as well as the duvet, and if the night is warm, to sleep under the sheet alone. Hotels often provide no top sheet, and I am sometimes reduced to sleeping under a bath towel. But in my own bed, I pull the sheet over me, shut my eyes and – rather like my cats – go out like a light.

Index

abrasives 41–2
acids 29, 46
acne 30
adding machines 120
Addis, William 36
Aircraft 51,106
alarm clocks 9–13, 17–18
alcohol 159, 163, 170, 172, 184
algebra 122–3
Ancient Egyptians 66, 108, 115, 157
Arabs, invention of soap 28–9
artificial light 18–22, 181–4
artists 112, 136, 141
ash 63–4, 106
Atkins, Anna 138–9
Atkins diet 159
Attenborough, Sir David 187

Babbage, Charles 120–121
bacteria 30, 39, 44, 45–6, 76–7, 175
bad breath 40–1
Bain, Sandy 127–8, 181
Baird, John Logie 185–6
balloons 96
Bank of England 116–17
Banting, William 159–60
barbers 48–50
batteries 9, 17–19, 38, 133
Beck, Henry 103–4
Bell, Alexander Graham 128–9
Bernoulli effect 26–7
bicycles 87–96, 104–5, 107
Big Ben 16
biological clocks 8–9
Biro, Laszlo 112
Black, Joseph 172–4
Blair, Tony 43
BMI (Body Mass Index) 161
body odour (BO) 45–6
Boolean algebra 122–3
Boulton & Watt 183
bowler hats 100–101
Boyle, Robert 15, 142, 182
Bramah, Joseph 82
breakfast 67–76, 162
Bristol Stool Form Scale 78–9
British Toilet Association 125, 187
Brunel, Isambard Kingdom 100, 169
brushes 33, 48–9

caffeine 25
camera obscura 137
cameras 134–6, 140–141, 149–50, 154
Canada 94–5
candles 12–13, 76, 181, 184
cannonballs 113–14
caoutchouc 96–7
carbon 19–20, 53, 112–13, 176
carbon copies 118, 120
carbonated drinks 170–1
cars 86, 88, 107
Caxton, William 117
Cayley, Sir George 106
cellulose 21, 55
cereals 67–71
chewing gum 41
China 115–17, 163–5
circadian rhythms 8–9
clay tablets 111, 115
cleaning 44
  clothes 57
  hair 32
  products 28–31
  skin 25–7, 45–6, 85
  teeth 36–43
Cleopatra 157
Clissett Wood, Herefordshire 62–5
clocks 15–16, 127, 135
  alarm 9–13, 17–18
  biological 8–9
  candle 12–13
  pendulum 13–17, 60
  water 11–13
clothes 54–60, 92–6
coal gas 182–4
Cochran (Cochrane), Josephine 180–1
coffee 25, 144, 145
commuting 102–3, 108–9, 126
computers 11, 108, 120–6, 132–3, 144
condensation 27, 48, 172–3
constipation 67–8
cooking 175–81
cotton 54–5, 59, 115
Crapper, Thomas 80
cuneiform 111
cyanotype process 138–9
cycling helmets 100–101

deafness 128–9, 159
Death Valley, California 131, 151
defecation 67–8, 76–9, 80
dehydration 44, 153
dental floss 39–40

depilation 50
detergents 28–30, 57, 179
Dew, Josie 40, 179
diets 158–61
digestion 69–70, 76, 175–6
digital photography 140–141, 143–4
dinner 174–81, 183–4
dishwashers 179–81
disposable razors/blades 51–3
documentaries 146–9, 150–153
drinks
  alcoholic 159, 163, 170–2, 184
  coffee 25, 144–5
  tea 26, 163–5
dry-cleaning 57
Dunlop, John Boyd 107

Earl Grey tea 163
earth closets 82, 84–5
Edison, Thomas Alva 19–21, 68, 118–19
eggs 72–4
Egypt 11, 136
electric light 18–22, 184
electric razors 46–7
electric toothbrushes 38–9
electricity 127, 184
Eliot, T. S. 124
elm 63, 65–66, 106
email 125, 131
energy 69–70, 71, 145, 172–3, 178
energy efficiency 22, 26, 87–8
evaporation 27, 32, 95
exercise 24–5, 88, 90, 159
eyes 9, 162

faeces 77–9
Faraday, Michael 96, 182
fax machines 126–8
Feynman, Richard 36
filaments 19–21, 37
film (photographic) 136, 139–41, 143
filming 151–3, 158
fish 162–163, 174
flashguns 134–6
fleeces 56, 95
'flow' 141–2
fluorescent lamps 21–2
fluoride 42
Flynn, Charlie 151–2
footwear 58–60
Ford, Henry 68
Fox Talbot, William Henry 139
friction 58–9, 128
Friction Lights 181
furniture 33–4, 62–7

Galileo Galilei 13–14, 152
gas lighting 183–4
gels 30–31, 47–9, 136
ghati (water clock) 12
Gilgamesh legend 111
Gillette, King Camp 51–2
Gillott, Joseph 111–12
glucose 69–70, 158
glycemic index 70–1
Goodyear, Charles 97
Gore-tex 95–6
graphite 112–15
gravity 15, 26
grease 28, 30, 57
Great Exhibition, Crystal Palace (1851) 82
Greeks
  optics 137
  shaving 50
  spoked wheels 106
  water clocks 11–12
green woodwork 33–4, 62–6
Greenwich 'pips' 9, 18
Grey, Charles, 2nd Earl 163
Gutenberg, Johannes 117

Hadrian's Wall 81
haemorrhoids 77
hair 30, 32–3, 46–53
hairdryers 32–3
Harrison, John 16
hats 99–101
al-Hazen 136–7
heat 166–8, 175–81
Holkham Hall, Leicestershire 100
hourglasses 13, 144
hydrogen 96, 183
hydrogen bonds 33, 55, 56, 57
hydrogen sulphide 183
hydrolysis 69–70, 176

ice 171–2
India 12, 92, 164
infrared radiation 136, 145
insulation 21, 93, 174
Internet 125–6
intestinal worms 85
iron 15, 67, 106
ironing 54–55

Japan 71–2, 128, 163
Jennings, George 82

Kellogg, John Harvey 67–8, 76–77
Kenya 69, 130, 153
klepshydra 11–12, 74
knots (nautical miles) 13
kohl 157
Ktesibios 11

laptop computers 108, 144
latent heat 172
latex 60, 96–7
lavatories 76, 79–80, 94
lead 42, 157–8
'lead' pencils 114
Leaning Tower of Pisa 152–3
leather 58, 67

lecturing 125, 168–9
Leitz, Gudrun 62, 65
light 134–5, 137
  artificial 18–22, 181–4
  photographic exposure 139–40
light bulbs 18–22
lipstick 156
Local Heroes 90–1, 169
London transport 102–4
lubricants 47–8
lunch 158–63

MacIntosh, Charles 97
make-up 43, 156–8
Marconi, Guglielmo 145
mascara 157
matches 181–2
Meso-Americans 60, 98
Mesopotamia 36, 105, 111
metals 15–16, 18–20, 42
Metropolitan Life tables 160–1
microwaves 145, 177–8
mirrors 34–5, 143
mobile phones 108–9, 122, 130, 132–3
Mohenjo-Daro 81
moisturisers 44
Moule, Reverend Henry 82, 84
mouthwash 41
Murdoch, William 183

NASA 77, 115
Newcomen engines 172–3, 177
nibs 111–12
'Nocturnal Remembrancer' 22–3
nylon 58–9

oak 64, 66
obesity 159–61
oil 48, 57
omega-3 174
opium wars 164
oscillation 9, 13–17, 38

palm computers 108, 126
paper 115–17
paper money 116–17
Papin, Denis 176
papyrus 111, 115
pay phones 130–1
pencils 112–15
pendulums 13–16
pens 111–12
perchloroethylene (perc) 57
PET (polyethylene terephthalate) 56
phone exchanges 131–2
photograms 137–8
photography 125–6, 134–44, 151
Pinchbeck Junior, Christopher 22–3
plague 39, 41
plasticisation 33, 55
pneumatic tyres 107
polyester 55–56
polymerisation 96
PopSwatches 60–1
porridge 68–9, 71
pressure cookers 176–7
Priestley, Joseph 90, 96
printing 117
promissory notes 116–17
PTFE (polytetrafluoroethylene) 95–6

Racetrack Playa 151
radio 145–9
radio signals 9, 18, 132–3
rain 98–9, *see also* waterproofing
razors 46–53
recycling 56
Romans
  artificial light 181
  barbers 49–50
  candles 12–13
  defecation 79
  footwear 58
  lavatories 81–2
  water organs 11
  writing 115
Royal Institution, London 19, 105, 125, 168–9, 187
Royal Society, London 15, 105, 138, 182, 187
rubber 60, 97, 107

satellite phones 133
Science Photo Library 143
SCN (suprachiasmatic nucleus) 8–9
shampoo 28–31
shaving 46–53
shoes 58–60
shower gel 30–31, 47
showering 25–8
Skara Brae, Orkney 66, 80–1
skating 128
skin 30, 44, 157
  cleaning 28–31, 45–6, 85
  moisturising 44–5
  shaving 46–53
sleep 9, 188
soap 28–9, 31, 47–8, 57
socks 58–9
sound recording 147, 150
spacecraft 77–8
steam 27, 33–4, 55
steam engines 172–3, 176–7
steel 53, 111–12
Stephenson, George 146
stimulants 25, 145
Stone Age 66, 81
Stringfellow, John 51
studios 148–9, 154–6
sulphur 30, 97, 182–4
Sumerians 36, 105, 111, 115
sun damage 44–5
sundials 10
surfactants 28–9, 30
Swan, Joseph Wilson 18–21
sweat 45–6, 56, 59, 95, 97
tea 25, 163–5

teapots 165–8
teeth 36–43
telegrams 126
telephones 108–9, 128–33
television 149–56, 184–7
telex 126
temperature 26–8, 94–5, 153, 171–2
texting 130
theobromine 25
time-keeping, *see* clocks; watches
toast/toasters 75–6, 159
Tomorrow's World 154–5, 185
tools 63–5
toothbrushes 36–9
toothpaste 41–3
transport
  bicycles 87–96, 104–5, 107
  cars 86, 88, 107
  London 102–5
  wheel development 105–8
tungsten 20–21
Turner, J. M. W. 112
Tyndale, William 69
typewriters 118–20
tyres 106–7

ultraviolet light 21, 44, 145
umbrellas 97–9
underground railways 102–5
urine 78, 85, 94

vacuum 19, 21, 173
vegetables 162, 174
Victoria, Queen 13, 100
vitamins 44, 162
Volta, Alessandro 18
vulcanisation 97, 107

Walker, John 181–2
Wallis, Trevor 98–9
watches 60–2, *see also* clocks
water 11, 25–8, 115, 135, 151–2, 164–5, 171–2, 176
water clocks 13
water closets 82
waterproofing 95–7, 99
Watt, James 172–3, 177
Wedgwood, Tom 137–8
Wellington boots 60
Westminster clock 16–17
wheel development 105–8
wood 114–15
woodwork 33–4, 62–6
wool 56–7, 59
word processors 122–4, 128
Wright brothers 106
writing 108–9, 111–20

PICTURE CREDITS

Jacket, contents pages and elsewhere Robert Wilson; pp19, 37, 64, 68, 69, 102–3, 131, 181, 183, 186 Mary Evans Picture Library; p14 metmuseum.org; p15 (Harrison), p79 (Crapper), p137 (Wedgwood), p142 (Boyle), p169, 172, 177 National Portrait Gallery; p16 National Maritime Museum; pp21, 29, 31 (hair), 33 (hair), 41, 44, 54, 55 (fibres), 56, 94, 113, 132, 133, 145, 158, 164, 174, 178, 185, 187 Science Photo Library; p49, p50 razorland.fr; p79 (Bristol Stool Scale) K.W. Heaton; p80 Thomas Crapper & Co; p81 (Roman fort) Martin Mortimore; p82 (Rev Henry Moule) Dorset County Museum; pp87, 91, 97 Dr Sue Blackmore; p88 IMechE Proceedings Archive; p96 Paul Bradshaw; p100 (SS *Great Britain*) Martin Chainey; p103 (underground map) London's Transport Museum © Transport for London; p123 Damodar Purohit; p124 PDPhoto.org; p142 Mustafa Hammuri; pp146, 147 Peter Siddoli; p152 Paul Bader; p153 (armour) Martin Mortimore, (Rift Valley) Karen Robinson; pp154–5 Patrick Titley; p159 www.shirleys-wellness-cafe.com; other pictures by Adam Hart-Davis.